This book tells the stories of scientists from Germany and other European countries who vanished during World War II. These erudite scholars contributed to diverse scientific fields and were associated with some of the world's leading universities and research institutions. Despite their proficiency, they all sought help from agencies to relocate to the UK in the 1930s, but were unable to secure the necessary assistance. The book highlights the extraordinary narratives of thirty such scientific refugees, delving into the reasons behind the unavailability of aid and presenting fresh insights into the tragic fates or astounding survival experiences of these individuals.

The Lost Scientists of World War II

David C Clary

University of Oxford, UK

The Lost Scientists of World War II

 **World Scientific**

NEW JERSEY · LONDON · SINGAPORE · BEIJING · SHANGHAI · HONG KONG · TAIPEI · CHENNAI · TOKYO

Published by

World Scientific Publishing Europe Ltd.

57 Shelton Street, Covent Garden, London WC2H 9HE

Head office: 5 Toh Tuck Link, Singapore 596224

USA office: 27 Warren Street, Suite 401-402, Hackensack, NJ 07601

Library of Congress Cataloging-in-Publication Data
Names: Clary, David C., author.
Title: The lost scientists of World War II / David C. Clary, University of Oxford, UK.
Other titles: Lost scientists of World War 2
Description: New Jersey : World scientific, 2024. | Includes bibliographical references and index.
Identifiers: LCCN 2023028658 | ISBN 9781800614758 (hardcover) |
   ISBN 9781800614918 (paperback) | ISBN 9781800614765 (ebook) |
   ISBN 9781800614772 (ebook other)
Subjects: LCSH: Jewish scientists--Europe--Biography. | Holocaust victims--Biography. |
   Holocaust survivor--Biography. | World War, 1939-1945--Science.
Classification: LCC Q128 .C53 2024 | DDC 509.2/2--dc23/eng/20230620
LC record available at https://lccn.loc.gov/2023028658

British Library Cataloguing-in-Publication Data
A catalogue record for this book is available from the British Library.

For any available supplementary material, please visit
https://www.worldscientific.com/worldscibooks/10.1142/Q0436#t=suppl

Desk Editors: Nimal Koliyat/Rosie Williamson/Shi Ying Koe

Typeset by Stallion Press
Email: enquiries@stallionpress.com

Printed in Singapore

About the Author

 **Sir David C Clary FRS** is Emeritus Professor of Chemistry at the University of Oxford, UK. He was President of Magdalen College, Oxford, from 2005–2020. He is an elected Fellow of many academies including the Royal Society, the Royal Society of Chemistry, the Institute of Physics, the American Physical Society, the American Association for the Advancement of Science, the American Academy of Arts and Sciences, and the International Academy of Quantum Molecular Science. He was President of the Faraday Division of the Royal Society of Chemistry from 2006–2008. From 2009–2013, he was the first Chief Scientific Adviser to the UK Foreign and Commonwealth Office. In 2016, he was knighted in the Birthday Honours of Queen Elizabeth II for services to international science.

Sir David is a theoretical chemist recognised for his pioneering work on the quantum dynamics of chemical reactions. He has published over 350 papers in this field. He was Editor of *Chemical Physics Letters* from 2000–2020 and a Reviewing Editor of *Science* from 2003–2016. He has won many prizes for his research including the Royal Society of Chemistry Meldola, Marlow, Corday–Morgan, Tilden, Polanyi, Chemical Dynamics, Liversidge and Spiers awards, and the Medal of the International Academy of Quantum Molecular Science. His book *Schrödinger in Oxford* was published by World Scientific in 2022.

Preface and Personal Acknowledgements

The idea for *The Lost Scientists of World War II* came from research I carried out on my previous book *Schrödinger in Oxford* (2022). While exploring Schrödinger's connections, I came across Fritz Duschinsky, a physical chemist from Czechoslovakia. He had published a paper on molecular spectra in 1937 that has become very influential, and then seemed to disappear. Intrigued, I delved deeper and searched various archives to uncover many new details about Duschinsky's life and career that have not been published before. I found that Duschinsky had applied to the British Academic Assistance Council in 1934 following the rise to power of Hitler and the Nazis. Despite helping many hundreds of other scientists to escape from countries in central Europe in the 1930s, this Council had been unable to assist Duschinsky in coming to the UK. I then discovered several other scientific refugees in a similar situation whose fates during the war were uncertain. The stories of these displaced scientists are told in this book.

Many of the lives of the scientists described in the book have tragic ends but some had good fortune, and others showed great courage and heroism. In all cases, the refugees were seeking earnestly to be able to continue their research careers in science during the extreme dangers of the 1930s and 1940s. I give examples right across the sciences and examine the trends in different subjects. The quote from an assistant to Winston Churchill that the Allies won the war "because our German scientists were better than their German scientists" has some truth.

The book shows, however, that there were several scientists from Germany, and also from other countries such as Austria and Czechoslovakia, who were unable for various reasons to come to the UK and make these contributions despite their scientific eminence. Biographical details of some of the scientists discussed here have been published previously and these have been brought together in this book to provide a comprehensive discussion and comparison of scientists from different subjects.

The archives of the Academic Assistance Council and the Society for Protection of Science and Learning are a unique source in the UK for biographical information, including reference letters from highly distinguished scientists that have been invaluable for describing the lives and careers of the individuals discussed in this book. New material for this book was also obtained from several other sources, archives and individuals in the UK and other countries.

I would like to give special thanks to descendants and colleagues of some of the scientists discussed in the book, including Professor Rosemary Pattenden (granddaughter of Alfred Byk), Michael Dreyfuss and Judy Navon-Dreyfuss (son and daughter of Paul Dreyfuss), and Professor Mansur Gilmullin (colleague of Alfred Lustig in Yelabuga) who provided personal copies of documents, information and photographs. I am also pleased to acknowledge helpful advice from Brian Collins (on Robert Eisler), James Deem (on Vladimir Lasareff and the National Belgian archives), Stefan Wolff (on Alfred Byk and Erich Lehmann) and Peter Voswinckel (on Hans Hirschfeld). I would also like to thank many archivists in several countries for their assistance, and details are given at the end of the book together with a list of copyright permissions. Several letters and quotes have been translated by me and are denoted in the text by (t).

Contents

About the Author   vii

Preface and Personal Acknowledgements   ix

Chapter One   Crisis in Europe in the 1930s   1

Chapter Two   Physics and Chemistry Non-Survivors   9
2.1 Friedrich Duschinsky   9
2.2 Emanuel Wasser   30
2.3 Marie Wreschner   44
2.4 Alfred Byk   57
2.5 Herbert Pese   66
2.6 Erich Lehmann   68

Chapter Three   Physics and Chemistry Survivors   73
3.1 Paul Dreyfuss   73
3.2 Three Chemical Physicists   80
    3.2.1 Vladimir Lasareff   82
    3.2.2 Paul Goldfinger   87
    3.2.3 Boris Rosen   92
3.3 Alfred Lustig   95
3.4 Giulio Bemporad   99
3.5 Karl-Heinrich Riewe, Part I   103
3.6 Friedrich Houtermans   105
3.7 Karl-Heinrich Riewe, Part II   120

Chapter Four   Top-Secret Refugees   **123**
4.1 Otto Frisch   123
4.2 Klaus Fuchs   126

Chapter Five   Refugees in Mathematics   **129**
5.1 Ludwig Berwald   130
5.2 Walter Fröhlich   133
5.3 Robert Remak   137
5.4 Otto Blumenthal   146

Chapter Six   Refugees in Medicine   **153**
6.1 Arthur Simons   154
6.2 Otto Sittig   157
6.3 Erich Aschenheim   160
6.4 Ferdinand Blumenthal   164
6.5 Hans Hirschfeld   171

Chapter Seven   Refugees in Biology   **181**
7.1 Hellmuth Simons   181
7.2 Vladimir Tchernavin   201

Chapter Eight   Refugees in Engineering   **215**
8.1 Alfred Rheinheimer   215

Chapter Nine   Refugees in Social Sciences   **219**
9.1 Robert Eisler   221
9.2 Paul Eppstein   231

Chapter Ten   Conclusions   **239**

References   245

Bibliography   273

Figures and Permissions   279

Permissions for Letters and Quotes   285

Index   287

CHAPTER ONE

Crisis in Europe in the 1930s

Monday, 30 January 1933 was a very cold day in Berlin, one that will be remembered forever. It was on this day that Paul von Hindenburg, President of the German Weimar Republic, made his fateful decision to appoint Adolf Hitler as Chancellor of Germany. This appointment set in motion a dramatic chain of events that would lead to World War II and a magnitude of death and destruction unequalled in history.

The Nazis under Hitler quickly established laws which would have appalling consequences. On 7 April 1933, the Law for the Restoration of the Professional Civil Service was introduced. This forbade "non-Aryans" and "politically unreliables" from holding positions in the civil service. A "non-Aryan" included someone whose parent or grandparent was of Jewish blood or religion. Academics with positions in universities and associated institutes of higher education in Germany were members of the civil service. Initially, the new law did not apply to people who were in posts at the start of the First World War, had served in the war or had lost a father or son in the war. However, this restriction was soon to be removed. The law was also extended to professionals such as doctors or lawyers. In addition, there followed the Nuremberg Laws which removed the human rights and citizenship of Jewish people and several other groups. The result was an exodus out of Germany of many of the most brilliant scientists and academics.

Some scientists, such as the leading physicist Max Born from Göttingen, who was from a Jewish family, realised at once that they would have to leave Germany with their families as soon as possible.[1] Several others, however, decided to stay expecting that Hitler would be removed

1

from power and the political situation would improve. Few anticipated the horror of the genocide that would be imposed by the Nazis.

Close observers in other countries were quick to realise the seriousness of the situation for many of the threatened scholars and scientists in Germany. Accordingly, assistance groups were set up to give advice and enable these academics to obtain suitable positions in other countries. In the UK and the USA, special agencies were set up with this aim while in other countries neighbouring Germany, such as Belgium, the Netherlands, France and Switzerland, leading professors worked to find places in their universities for academic colleagues facing difficulty in Germany.

The Academic Assistance Council (AAC) was established in the UK in 1933 to assist university teachers and researchers in all subjects who had been forced out of Nazi Germany.[2] Following a suggestion from the Hungarian physicist Leo Szilard, the AAC was set up by the economist and liberal politician William Beveridge with the support of many distinguished members of the British establishment. The Council he established included leading UK scientists such as the Nobel Laureates Lord Rutherford (as President), William H. Bragg, Lord Rayleigh, J. J. Thomson and Charles Sherrington, together with other prominent academics and intellectuals in the social sciences and humanities including J. Maynard Keynes, Gilbert Murray, G. M. Trevelyan and A. E. Housman. The honorary secretary was Charles S. Gibson and the full-time general secretary was Walter Adams. The assistant secretary was Esther (Tess) Simpson who became the closest correspondent of the academic refugees and the name best remembered by them.[3] The initial establishment of the AAC allowed for a fund to be set up through donations from individuals and institutions that proved to be vital in assisting scholars.

It was realised at once by the AAC that it would have an important role in liaising with the British Home Office to obtain suitable visas and work permits to allow the scholars from Germany to come and work in the UK. As is also true in the present day, there was some vociferous opposition to immigrants taking up jobs that would have normally gone to British nationals. Accordingly, a careful case had to be made that an immigrant scholar would have a special expertise that was not available in the UK and was valuable to the future of the country. In addition, very few new senior posts, such as professorships, could be created due to the parlous financial state of many British universities in the 1930s. This meant it was very difficult to assimilate highly distinguished scholars. For example, Erwin Schrödinger came to Magdalen College, Oxford, from Berlin in November 1933, but a permanent chair could not be found for him and he returned to his home country of Austria in 1936.[4] Similarly, Max Born came to Cambridge from Göttingen to take up a temporary lectureship at the University of Cambridge in 1933 that only had a tenure of three years.[1]

Very soon after Hitler came to power in 1933, the AAC began to receive letters from scholars who had been expelled from their positions at German universities and were hopeful of taking up a position, even a temporary one, in the UK.[2] The procedure was that a form was then sent to the scholar to provide the full details of their academic history and personal situation. A list of referees who might support their case was needed as was a full publication list. There were other questions asked such as the religion and family members of the scholar, which countries they would be prepared to settle in and personal financial details.

The referees listed by the scholar were particularly crucial and it was assumed the more esteemed the referee, the better the opportunity for finding a placement. Thus, world-famous Nobel Prize winners such as Albert Einstein (for physicists and refugees in several other subjects) and Fritz Haber and Walther Nernst (for chemists) were often used as referees while world-leading mathematicians such as G. H. Hardy and David Hilbert were also often chosen. Each case was then dealt with first by the AAC assistant secretary Tess Simpson, consulting with the general secretary, to see if an academic place was possible. She would send the file of the scholar to the leading professors in the field of the applicant at the top universities in the UK to see if, in the first instance, a temporary placement could be found with some minimal financial support. If this proved possible, then the AAC executive committee would try to award an additional small grant to assist the scholar. This would normally provide adequate evidence to support the case for a working visa to be issued. Tess Simpson also corresponded frequently with the scholar informing them of the progress made and often asking for further information. The detailed files built up on each scholar have been archived comprehensively and are available in the Bodleian Library at the University of Oxford.[5] They have proved to be a unique and hugely valuable resource for scholars studying academic refugees in the period building up to World War II.[6–8]

In 1936, the AAC expanded its remit to deal with persecution in other countries, established a new trust fund and was renamed the Society for the Protection of Science and Learning (SPSL). Initially, the AAC/ SPSL had considerable success in placing refugee scholars from Germany.[9] However, as time moved on, the number of suitable academic places diminished as did the available funds. Then, in 1938, an increasing number of scholars from countries other than Germany sought assistance. Following the Anschluss in March 1938, several applications came from scholars based in Austria where the anti-Semitic laws of the Nazis were introduced at speed. After the decisions made in September 1938 in the Munich Agreement between Germany, Great Britain, France and Italy, applications to the SPSL then came also from Czechoslovakia. With Vienna and Prague being particularly active scientific centres, this presented a major challenge to the SPSL to place highly distinguished academic refugees from these cities who often worked in similar fields. The SPSL also received a small number of applications from academics in other countries such as Italy and Poland, and even from the Soviet Union.[5]

The AAC/SPSL had extraordinary success in bringing many refugee academics to the UK who then had outstanding academic careers in their newly adopted country.[9] Assistance was given to over 1,000 scholars in the 1930s of whom as many as sixteen went on to win the Nobel Prize and eighteen were knighted. Over one hundred became Fellows of the Royal Society or the British Academy.[9] It is no exaggeration to say that the lives of many of these scholars were saved by the AAC/ SPSL. The contributions of these scholars, and their children in several cases, to the scientific, intellectual and cultural life of the UK cannot be overstated.

The importance of the exodus from Germany of so many top scientists was not fully appreciated by the Nazis. Carl Bosch and

Fritz Haber had invented an efficient catalytic method to produce ammonia, which was vital for fertilisers, chemicals and munitions, and had been involved in creating the highly profitable company I. G. Farben. Haber was one of the first famous scientists to be forced out of Germany and, in a personal meeting with Hitler, Bosch complained bitterly that the expulsion of so many Jewish physicists and chemists would put research in German universities back by a hundred years. Hitler replied, "Then we will work a hundred years without physics and chemistry."[10] In due course, the contributions of a number of refugee scientists to the UK and USA war efforts were highly significant. Sir Ian Jacob, Winston Churchill's military secretary, once said to the Prime Minister that the Allies won the war "because our German scientists were better than their German scientists."[11]

The AAC/SPSL also communicated regularly with a sister body in the USA, the Emergency Committee in Aid of Displaced Foreign Scholars, based in New York.[12] When the possibility of a placement looked doubtful in the UK, the papers of the refugee were often sent on to the Emergency Committee. The procedure normally used by the Emergency Committee was to identify a university position that was available, put together a gathered field in the area of speciality and then send on the details to the university. This was a competitive process which was different from the individual basis that was used by the AAC/SPSL.[13] Nevertheless, the Emergency Committee managed to find placements for over 300 applicants and some of these were scholars who were already world famous, such as Thomas Mann and Herbert Marcuse, and also future Nobelists such as Konrad Bloch and Max Delbrück. However, the records of the Emergency Committee show that support could not be obtained for over 5,000 applicants.[12] Nevertheless, several of the most brilliant of these scholars, such as John von Neumann, Hans Bethe and Fritz London, still managed to obtain appointments by direct applications to US universities or institutions.[14-16] In addition, some world-famous academics such as Albert Einstein had been appointed directly to prestigious posts in the USA just before the Emergency Committee was established. Furthermore, some of the most promising refugee scientists, such as Eugene Wigner, managed to obtain posts with US universities without applying to the Emergency Committee.[17]

There were also other bodies which helped refugees, including the Notgemeinschaft Deutscher Wissenschaftler im Ausland, which was based initially in Switzerland, and the British Federation of University Women.[18–20] Several endowed institutions provided significant funding for scientific refugees such as the Rockefeller Foundation in the USA and the Fondation Francqui in Belgium.[13] Furthermore, some major industrial companies, such as the UK Imperial Chemical Industries, also supported refugee researchers. All of these bodies communicated closely with the AAC/SPSL.

The success of the AAC/SPSL is deservedly acclaimed but not every academic who sought help in the 1930s could be assisted. A small number failed to leave their home countries or other occupied territories. During and after World War II, a major effort was made by the secretariat to track down those academics in the AAC/SPSL records whose whereabouts were unknown. It was found in some of these cases that the individuals had managed to live through the war in their own or occupied countries, or had found safe appointments elsewhere. However, a small number were found not to have survived or just disappeared.

In this book, we describe the stories of these "Lost Scientists of World War II." We consider individuals across the sciences, including chemists, physicists, biologists, engineers, mathematicians, medical scientists and social scientists. We discuss the special reasons why assistance could not be provided in these cases despite extensive efforts from the AAC/SPSL and other institutions and individuals. We also compare and contrast the challenges of the different subject areas. For several individuals, we have discovered new information on their wartime activities and fates. A small number survived the war without the SPSL ever realising what happened to them. Several of the individuals had family members to care for, including spouses, children and parents, which made their situation more complicated. In some cases, we also describe the stories of these family members. In examining thirty refugee scientists in detail, we have found that many of the individuals had tragic fates but there are also examples of good fortune, heroism and great courage. Several of these scientists made major discoveries in their research that have lasted to the present day. The full details of their lives deserve to be told.

CHAPTER TWO

Physics and Chemistry Non-Survivors

Physics and chemistry were well-developed academic subjects in Germany and neighbouring countries such as Austria, Switzerland and Czechoslovakia in the early 1930s. This was due to the great breakthroughs made in physics in the early part of the 1900s in those countries and also because of the importance of the rapidly expanding chemical industries.[21] Accordingly, when the expulsion of academics from Germany started in 1933 and from Austria and Czechoslovakia in 1938, there were a large number of applicants in physics and chemistry to the AAC (before 1936) and the SPSL (from 1936). However, there were several unsuccessful applicants whose research was very much on the border between chemistry and physics in interdisciplinary areas, which are nowadays classified as chemical physics or physical chemistry. In terms of academic positions and funding, interdisciplinary areas have always presented problems even in the present day. Therefore, in our analysis in this chapter, we discuss as a group examples of refugees classified as both physicists and chemists.

2.1 Friedrich Duschinsky

Friedrich (Fritz) Duschinsky (1907–1942) was from Czechoslovakia. In 1937, while working in Russia, he published a paper describing a method for using quantum mechanics to calculate molecular spectra that has

Fig. 2.1   Friedrich Duschinsky in his Belgian Aliens Registration Certificate of 1934.

become highly cited and continues to be widely applied in the present day.

Duschinsky was born in the town of Gablonz on 26 February 1907 (his first name was given in various records as Fritz, Friedrich, Frédérique, Frederic, Franz or Bedrich). This town is situated 88 km northeast of Prague and is close to the modern borders of Germany and Poland in the Sudeten region of the former Czechoslovakia. Gablonz was in the Austrian–Hungarian empire until 1918, when it became part of Czechoslovakia and had a population of about 40,000 people. Gablonz was the name used by inhabitants who spoke German and the town was also called Jablonec nad Nisou by Czech speakers, which is the modern name. The town became prosperous with the production of glass and artificial jewellery being a major industry.[22]

Duschinsky came from a Jewish family and was a prodigy in physics. His father Alexander ran a factory involved in the manufacturing of artificial pearls. Alexander was in the Austro-Hungarian Army in World War I. The family spoke German at home and Alexander Duschinsky was a staunch supporter of Germany until the late 1930s.[23] Fritz Duschinsky's mother's name was Jenny (née Strenitz). Fritz went to a German-speaking grammar school in Gablonz where there were few Jewish students.[23] At that time, there was a small amount of anti-Semitism but nothing compared to what would develop in the 1930s.[23]

His higher education was initially at the German University in Prague in 1926 and then at the Sorbonne in Paris from 1927–1928 where he studied under Marie Curie.[24] After this, he moved to the University of Berlin for his Dr. phil. under the supervision of Peter Pringsheim, a world-leading expert on molecular luminescence and phosphorescence. Pringsheim had interacted and corresponded at length with Albert Einstein and James Franck, and was the first to use the phrase "the Raman effect."[25] Duschinsky's research thesis was on the experiment and theory of the absorption and emission of radiation by atoms and molecules. In Berlin at that time, many of the most celebrated scientists were present whose fundamental research was defining modern physical science, including Albert Einstein, Max Planck, Erwin Schrödinger, Max von Laue, Walther Nernst, Lise Meitner, Fritz Haber and Otto Hahn. As a research student, Duschinsky was able to attend lectures and seminars alongside these great names who freely discussed science with him.

In 1933, Duschinsky published three single-author papers in the *Zeitschrift für Physik* on "The influence of collisions on the lifetime of excited Na atoms," "The temporal course of the intensity of intermittently excited resonance radiation" and "A general theory of experimental arrangements (fluorometers) for measuring very short light durations."[26–28] In recent years, the theory developed in the last two of these papers has been influential in the development of a widely used multifrequency phase modulation fluorometry technique for measuring molecular fluorescence.[29]

In these papers, he showed himself adept at both theory and experimentation on the interaction of radiation with molecules. Duschinsky's doctoral thesis was classified as "magna cum laude" and it looked as though he was set for a stellar career in research on the borders between physics and chemistry. Following his successful thesis defence, he was appointed on 1 January 1933 as a research assistant to Professor Karl Weissenberg, an expert on crystallography at the Kaiser-Wilhelm-Institut for Physics in Berlin.

The acknowledgements for two of Duschinsky's single-author papers indicated a close interaction with Erwin Schrödinger, who had taken Max Planck's professorship in Berlin and was thanked for "friendly advice, and valuable and demanding criticism."[26,27] On the announcement of Schrödinger's Nobel Prize for Physics, following his move from Berlin to Magdalen College, Oxford, in November 1933, Duschinsky wrote a letter of congratulation by hand from Gablonz stating, "As a former enthusiastic listener, I express my great joy on the award of your Nobel" (Figure 2.2).[30(t)] His listening was to prove useful as Duschinsky was to soon publish an application of Schrödinger's wave mechanics to the calculation of spectra of polyatomic molecules which would make his name well known in chemical physics in the present day.

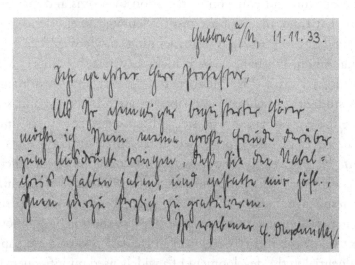

Fig. 2.2   Letter from Duschinsky congratulating his teacher Schrödinger on his Nobel Prize, written from Gablonz on 11 November 1933.[30]

Fig. 2.3   Peter and Emilia Pringsheim in their Belgian Aliens Registration Certificates of 1933.

However, as was the case for all the Jewish people in Germany, everything changed on 30 January 1933 when Adolf Hitler was appointed Chancellor of Germany. Pringsheim, who was born in Munich, was from a Jewish family who had made a fortune from railways and coal mining, and this made him particularly vulnerable to the Nazis. His wife Emilia was Belgian and they moved quickly in September 1933 to Brussels (see Figure 2.3).[31] There, Pringsheim took up a position at the Free University of Brussels arranged by the head of the department of physics, Professor Auguste Piccard. Piccard was internationally famous as he had undertaken high-altitude balloon ascents during which he became the first person to enter the stratosphere. Born in Switzerland, he was a well-known and flamboyant character in Brussels. He was the inspiration for the iconic character Professor Cuthbert Calculus in the *Adventures of Tintin* created by the Belgian cartoonist Hergé and known to children the world over even in the present day.[32] By being a local participant at the famous Solvay conferences of the 1920s and 1930s in Brussels, the well-connected Piccard had also got to know the leading physicists of the time and even the Belgian royalty.

Duschinsky was dismissed from his position in Berlin on 1 April 1933 under the new laws forbidding the employment of Jews by the Civil Service.[24] His first thought, following advice from Weissenberg, was to try to find a position in the UK or the USA and he wrote two letters from Gablonz on 5 July 1933. One letter was sent to Professor Abraham

Flexner, the founder of the Institute for Advanced Study (IAS) at Princeton.[33] In this Institute, positions had been found for several leading figures in Theoretical Physics and Mathematics from Germany including Einstein and Hermann Weyl. Duschinsky was too junior to be of interest to Flexner (and even Schrödinger had visited IAS in 1934 hoping for a position there, but an offer was not forthcoming).[4] Flexner, however, did pass on Duschinsky's application to the Emergency Committee in New York. The Assistant Secretary Edward R. Murrow then wrote on 14 August 1933 to inform Duschinsky, "We shall be very happy to keep you in mind and shall communicate with you at once should we hear of any suitable openings in this country, although it is somewhat difficult to be optimistic as to results."[33] Murrow himself was to achieve celebrity status by becoming the leading CBS radio reporter reporting back to the USA on the dramatic developments in Europe in the 1930s and 1940s. During the Blitz in London in 1940, his starting catchphrase "This is London" and finishing comment "Good night and good luck" became his famous trademarks.

The second letter from Duschinsky, sent also on 5 July 1933 from his address Schillergasse 8, Gablonz, was to Professor Frederick Donnan, a physical chemist at University College London.[24] Donnan had already interacted with the AAC in finding placements for displaced scholars from Germany and he forwarded Duschinsky's papers to the AAC.

Duschinsky then followed his former research supervisor Pringsheim to the Free University of Brussels on 16 January 1934 as a research assistant under the overall supervision of Professor Auguste Piccard (Figure 2.1).[34] It seemed at the time that this was a sensible move to enable him to continue his productive research on atomic and molecular spectroscopy outside the sphere of the Nazis. In Brussels, Duschinsky published with Pringsheim on the influence of foreign gases on the observation of the fluorescence of iodine vapour.[35] Experimental results were discussed and explained in terms of potential energy curves. This analysis was quite advanced for its time and would form the basis for a subsequent and highly influential paper from Duschinsky on a theory for calculating the electronic spectra of polyatomic molecules.

Duschinsky's appointment in Brussels, however, was only until May 1935 and he wrote directly to the AAC on 3 March 1934. A detailed

AAC form was then sent on to him requesting more information, and this was received back in London on 13 September 1934. In order to indicate his experience in experimental optical physics, Duschinsky stated, "I would not accept commercial, but any industrial position of technical nature, preferably development research."[24] He also wrote that his expertise was in "Fluorescence, molecular spectra, general physical optics, high frequency techniques." He stated that he was Jewish and would be prepared to work in tropical countries "if free of malaria" and also the Far East, Soviet Russia and South America. He was not married and this, in principle, made him more flexible to move between countries than some of the other scientists discussed in this book. However, he somewhat naively wrote in a subsequent letter: "Concerning the opening as teacher on which you informed me, please do not submit my name; my English knowledges are so bad that I cannot engage to such employment."[24] This statement would not have encouraged the leading professors in the UK to offer a place in their laboratory where speaking and writing in good English was essential.

Duschinsky's referees were Peter Pringsheim, Karl Weissenberg and Walther Nernst who, by any standards, were an impressive set of supporters. Pringsheim's reference for Duschinsky stated in very strong terms:

> He has acquired great experimental skills and, in addition, he also completely independently gave the method, which had previously been used in various ways, the correct theoretical basis through thorough mathematical discussion and defined its effectiveness. He published the results obtained in this way in two special works. During my teaching career I have only met a few students who took such a lively interest in all scientific questions that arose, but who also contributed so much to their solution through their own thinking.[24(t)]

Nernst had won the Nobel Prize for Chemistry in 1920 for work that led to the Third Law of Thermodynamics and was particularly highly regarded in the UK. His more general reference stated:

> Kind regards, Mr. Duschinsky! In response to your inquiry the day before yesterday, I am happy to confirm that I got to know you better over the course of several years in the physics institute of the University

of Berlin, which I head, and in which I found plenty of opportunity to seek your knowledge, get to know you and appreciate your experimental skills. My best wishes accompany you on your further progress. With best regards, W. Nernst.[24(t)]

On receipt of a full application and references, the secretary of the AAC, Tess Simpson, followed her well-defined procedure of enquiring with the leading professors in the UK universities if there may be space in their laboratories to accommodate a new researcher on a temporary basis. Some kind of employment was essential to enable an academic to have permission to enter the UK. This was a firm requirement from the Home Office, with whom the AAC was very careful to keep on good terms. The AAC itself would also sometimes award a small additional grant from a small fund it had built up from numerous donations, and decisions on this were made by an expert Executive Committee.

Following the new laws implemented by the National Socialists, Max Born had moved in 1933 from Göttingen in Germany to a temporary position at the University of Cambridge.[1] In Göttingen, he had built up a large and very eminent research group in theoretical physics which had included researchers who were to become some of the greatest names in 20th century physics, including Werner Heisenberg, Wolfgang Pauli and Enrico Fermi. Accordingly, Born was often consulted by the AAC on applications from physicists and, in this case, he was asked for his opinion on Duschinsky. However, Duschinsky's single-author papers in the *Zeitschrift für Physik* had only just been published and Born had been much distracted by the huge disruption to science in Germany and his own move to England in 1933. On Duschinsky, Born wrote, "I don't know his work and cannot say anything about him, you should write to Pringsheim."[24] As Duschinsky's papers were on the borders between chemistry and physics, they had not been read at that time by many physicists in Germany or the UK who were concerned with more fundamental questions such as those of atomic and nuclear physics.

It is unfortunate that Duschinsky's application was not sent to John Lennard-Jones who had just been appointed in 1932, at Cambridge University, to the first ever Chair in theoretical chemistry and had pioneered molecular orbital theory for interpreting molecular spectra. He had been one of the organisers of a Faraday Discussion meeting on Molecular Spectra and Molecular Structure in Bristol in 1929.[36] Thus, the best links for Duschinsky in the UK had not been made and no firm offers came his way. Tess Simpson did forward information diligently to Duschinsky on several possible appointments outside the UK including in New Zealand, Quito in Ecuador, and Ghent in Belgium. He informed the AAC that he applied for these positions, but there is no evidence that he was considered further. In his extensive correspondence with the AAC, he always signed his letters F. Duschinsky, so the AAC was not aware of his first name and erroneously used "Franz" in its records.

Tess Simpson had also been informed by the Emergency Committee in New York of a position for a post at Wesleyan College, Connecticut, and she sent on the full set of Duschinsky's papers to the Emergency Committee.[33] This Committee had taken advice from Dr. Fritz Demuth of the Notgemeinschaft Deutscher Wissenschaftler im Ausland. This organisation, based at that time in Zurich, had been set up by academic refugees to assist scientists who had been expelled from Germany. It had close interactions with the AAC. Demuth stated, in connection to a possibility of a chemistry position in Venezuela, that "Duschinsky was not a chemist."[33] His interdisciplinary area of molecular spectroscopy research fell between physics and chemistry and this did not help him in finding a suitable academic post.

In November 1934, Rudolf Peierls, who had close connections with Russian scientists, wrote to the AAC with information on a position for a physical chemist, which may have been appropriate for Duschinsky, at the Institute of Experimental Biology in Leningrad, of which Alexander Gurwitsch was the director.[24] The requirement was for a physicist who could explain some of the unusual effects they had observed through the interaction of light with biological systems. By this time, however, Duschinsky was already thinking of a position at another institute in that city in Russia. Sergei Vavilov and Alexander Terenin had both studied with Pringsheim and had then gone on to lead the National Institute of Optical Sciences in Leningrad. Vavilov and Terenin were impressed with Duschinsky's papers and he was offered an attractive position which had the rank of university professor and which would enable him to work independently in Leningrad doing both theoretical and experimental research.[37] Despite the political cloud over the Soviet Union, Vavilov was an influential member of the Soviet Academy of Sciences, of which he would eventually become president.[38]

The Leningrad Institute had an excellent scientific reputation and Duschinsky started there in April 1936. In one year, he published three papers in Russian journals.[39–41] He wrote his last and most influential paper "The interpretation of the electronic spectrum of polyatomic molecules. I. On the Franck-Condon principle," which he submitted to *Acta Physicochimica URSS* on 19 July 1937.[41] The paper carefully applied Schrödinger's wave mechanics to the overlap of the vibrational motions of ground and excited electronic states, and his simplified mathematical analysis using matrices provided a general procedure for calculating the spectra of polyatomic molecules in the ultraviolet and visible regions. His paper acknowledged V. Fock, A. Terenin and M. Eliaschewitsch, who were all then working in Leningrad. Vladimir Fock gave his name to the widely used Hartree–Fock theory of electronic structure calculations. Michael Eliaschewitsch (1908–1996), also called Eliashevich, published fundamental work on the theory of molecular vibrations.[42] He was still going strong in 1993, when the author of this book met him at a conference in Minsk, Belarus.

A major international congress was organised on *Luminescence* in Warsaw in 1936 and was presided over by Pringsheim. Duschinsky was chosen to attend as one of a seven-member delegation from the Soviet Academy of Sciences. In Duschinsky's own words, "Although all members of the delegation had registered their participation and the papers had already been printed, the (Soviet authorities) banned their exit at the last moment and this made participation in the congress impossible."[37] This was a great disappointment as attendance at the meeting, and the opportunity to communicate his new ideas on molecular spectra, could have firmly established an international reputation for Duschinsky.

By 1937, foreign scientists working in the Soviet Union were coming under suspicion and there were several arrests. This included Fritz Houtermans who is discussed in a later chapter in this book. Just after his last scientific paper was published, Duschinsky was informed in September 1937 that he would have to become a citizen of the USSR if he wished to continue working in Leningrad.[37] This was unacceptable to him and he returned in October 1937 to his home town of Gablonz.

In Gablonz, there was no university for Duschinsky to continue his very promising research, but he put his experimental abilities to good use through employment as an optical engineer with a local company J. Altmann, making specialist optical items.[37] His statement to the AAC that he had the ability and experience to accept "any industrial position of technical nature, preferably development research" was becoming highly relevant.[24] One of the family members working for J. Altmann was Maximilian (Max) Altmann whose wife Hilda had published several patents on using glass in a sophisticated way to design reflectors with carefully calculated geometrical cavities for use on bicycles, automobiles and similar applications.[43] Her patents were published from the address Reichenbergstrasse 37, Gablonz. Her innovative designs continue to be adapted in many new patents to the present day.

In September of 1938, the political situation in Czechoslovakia was becoming very tense. With a majority of German speakers residing in the north part of the country, including Gablonz, Hitler was threatening to invade Czechoslovakia just as he had invaded Austria some six months earlier. This led to a mobilisation of the army in Czechoslovakia. Erich Duschinsky, Fritz's brother, had worked in a company that manufactured imitation jewellery run by an aunt in Gablonz, which had a successful export business. He gave an extensive interview in 1988 on his life up to the end of World War II.[23] Born in 1911, Erich was four years younger than his brother Fritz. He was a Zionist activist in Gablonz and in 1938 developed an enterprising scheme to bring a group of Jewish children from Czechoslovakia to Palestine. He was able to travel to London from Scandinavia to attempt to make the arrangements for this in September 1938 when he was also doing work as part of the artificial jewellery export business. After London, he planned to travel to Marseilles in France to arrange a boat to take the children to Palestine.

At this time, Erich Duschinsky also attempted to assist his brother Fritz to move to London where they could work together on the manufacture of artificial jewellery. Accordingly, Fritz travelled to Brussels and then took a plane to Croydon airport near London in September 1938 where his brother was waiting for him. However, the immigration authorities in Croydon did not give permission for Fritz to enter the country. The reason they gave was the mobilisation of troops in Czechoslovakia and Fritz was of conscription age. Erich, however, felt the real reason was the evidence in Fritz's passport that he had worked in Leningrad.[23] Ever since the Russian Revolution, the British government had always had a negative view of the Soviets. Accordingly, Erich commented that he was only allowed a short meeting with Fritz at the airport who then took a plane back to Brussels and moved on to Paris to join an "old friend."[23] This was the last time that Erich saw his brother Fritz.

If the immigration officer at Croydon airport had made a different snap decision, then Fritz Duschinsky would have been able to spend the years of the war in the UK and his life would have been saved. There is no evidence in Fritz Duschinsky's police file that he attempted to stay in Brussels and undertake research work again with his mentor Pringsheim in 1938. Because of the mobilisation of troops in Czechoslovakia, Erich Duschinsky was unable to take the children to Palestine, although he did subsequently work for Jewish agencies in assisting refugee children on the east coast of Britain in Lowestoft and Harwich.[23]

Following the Munich Agreement of 30 September 1938, Germany annexed the Sudetenland which included Gablonz.[22] On 10 November 1938, the magnificent synagogue in Gablonz, designed by the distinguished architect Wilhelm Stiassny and built in 1891, was burnt to the ground in the Nazi's Kristallnacht.[44] At that time, over 800 Jewish people were living in Gablonz and many were attacked or taken into custody. There was an exodus from the town. Some moved to stay with friends living in less dangerous places in the region while others escaped from Czechoslovakia to what at the time appeared to be safer countries. Fritz and Erich Duschinsky had already had the foresight to leave Czechoslovakia in September 1938.

Fritz Duschinsky moved to the Paris region of France and was joined by Max and Hilda Altmann and their teenage son Alexandre.

Erich Duschinsky commented that Fritz was involved in the war effort either for the French government or the Czech government in exile.[23] Duschinsky then published, with the Altmanns, as many as six innovative patents in 1939 on a range of topics. His first such publication with Max Altmann was received on 1 April 1939 from the Seine (Parisian) region of France. Published in French, it described the invention of an "apparatus for navigation at night or in foggy weather."[45] On August 22, 1939, just two weeks before the outbreak of World War II, he published another patent, with Hilda Altmann, on an auto-collimating system.[46]

Initially, these patents from Duschinsky and his co-workers were on optical inventions but, after the start of the war on 3 September 1939, they became more relevant to devices needed for the armed conflict. This included a patent on the detection of moving objects such as aeroplanes and even a proposal for a new gas mask.[47,48] This was published on 16 October 1939 with Siegwart Hermann, a well-established inventor who was also a refugee from Czechoslovakia in Paris and was, for a short period in 1939–1940, appointed as an "economic advisor to the Czechoslovak Government in Exile," which was based in Paris for a brief episode before moving to London.[49,50] Two days later, Max Altmann, Duschinsky and Hermann proposed "a system which makes it possible to direct a large part of luminescent light towards the source of exciting light," a subject linked to Duschinsky's fundamental research work with Pringsheim in Berlin and Brussels.[48] On 1 November, with Hermann and Max Altmann, another patent was published "on the use of coloured lights for signalling in the time of war."[51] Then, the final patent from Duschinsky with the same authors was published on 21 November 1939 on the enhancement of fluorescent flashlights as used in troop movements.[52] After that, Duschinsky's patents stopped. However, Hilda Altmann published a patent on a reflector apparatus device, dated 1 February 1941 from the Seine-et-Marne district, with the Société d'Exploitation des Verreries de Bagneaux et Appert Frères Réunie.[53] A similar patent was also submitted by the same Société on 21 October 1942, but without Hilda Altmann as a contributor.[54]

The implication of this patent of 1 February 1941 is that the Altmanns, and most probably their close collaborator Duschinsky, had moved to Bagneaux-sur-Loing following the invasion of France and takeover of Paris by the German forces in June 1940. Bagneaux-sur-Loing is a small town, which then had just over 1,000 inhabitants, and is situated in the Seine-et-Marne region of France, 88 km southeast of Paris. Like Gablonz, it was well known for its glass industry and had become the centre in France for the new glass material Pyrex in the 1930s.[55] After the capitulation of France, Bagneaux-sur-Loing was situated in the German-occupied zone and it is unclear why the group from Gablonz did not move further south into the Vichy zone, which at that time may have been somewhat safer.

The close collaborator of Duschinsky and the Altmanns in Paris, the highly innovative Siegwart Hermann (1886–1956) from Prague, was already famous due to his invention of a fermented drink now widely known and distributed as "Kombucha" (see Figure 2.4).[56,57] He was well connected and he managed to leave France with his wife Erna and daughter Marianne, taking the boat *SS Magallanes* from Lisbon and arriving in the USA on 28 January 1941.[58] There, he continued with his patents, including one on 10 May on a "system and device for signalling purposes," which was closely connected to the work of Duschinsky and

Fig. 2.4   Siegwart Hermann, inventor of Kombucha (~1914).

the Altmanns.[59] Hermann also reproduced as a single author in the USA on 23 May 1941, and without acknowledgement to Duschinsky and Max Altmann, the gas mask patent these three had co-authored in France.[47,60] This patent from Hermann has subsequently been cited several times in more recent works involved with the patenting of breathing apparatus. The gas mask patent first published in France on 16 October 1939 has had no citations. The versatile Hermann subsequently had a successful career in the USA, mainly involved with the chemistry of fermentation linked to his invention of Kombucha.[49]

During this period, the Emergency Committee in New York was again considering Duschinsky. A letter was sent from Betty Drury, the Assistant Secretary, on 7 March 1940 to George Warren of the President's Advisory Committee on Refugees with a list of academics from Europe still on its books who could possibly be suitable for a new initiative in the Dominican Republic.[33] This country was finding space for Jewish people following encouragement from the USA. However, there is no evidence that an attempt was made for a further communication on this topic with Duschinsky and, by this time, it is very unlikely he could have been contacted.

The French Resistance was active around Bagneaux-sur-Loing and there were several arrests by the Gestapo in that region in 1942.[61] On 21 October 1942, there was a major roundup of Jewish people in the Seine-et-Marne region of France and this included Bagneaux-sur-Loing. The three members of the Altmann family and Duschinsky were arrested on that day by the Gestapo.[62] A post-war letter from Duschinsky's brother Erich to the SPSL mentioned that Fritz had been "denounced".[24] The Nazis and the cooperating French authorities kept detailed lists of their prisoners from which it is possible to establish the next movements of the four people from Gablonz. They were initially placed in the Melun prison on the main road from Bagneaux-sur-Loing to Paris.[63] Then, they were transferred to the notorious internment camp of Drancy in Paris and were imprisoned there on the second staircase. Fritz Duschinsky's name was recorded as "Buchinsky, Bedrich" in the typed records for transportation from Drancy but with the correct birthdate of 26 February 1907, birthplace as "Gablontz" and current address Bagneaux S/Loing (see Figure 2.5).[64] His nationality was undetermined. Bedrich is the old

```
Drancy-escalier 2                    -4-                                    21

DIAMANT Jetty              24.11.91 Seret        52 rue de la Goutte d'Or
née RACHMITH              roumaine               tailleur

DORFMAN Edla              31.10.69 Wegrow        60 Avenue des Pâquerettes
n"e GARBOS                Polonaise              S.P.        MONTFERMEIL

DORMMAN Leib              1.9.70 Wegrow          50 Avenue des Pâquerettes
                         polonaise              S.P.        MONTFERMEIL

BRUKER Ida               18.12.70 Lejuibi       19 rue de Verdun, BIARRITZ
                         Allemande              S.P.

BUCHINSKY Bedrich        26.2.07 Gablontz       BAGNEUX S/LOING
                         Indet.                 S.P.

K    dit EBE Anchel      22.12.77 Varsovie      228 rue St-Denis
                         ref.russe              Marcquinier

EXHHAJZEN Zwa            1906 Adamon            263 Fbg. St-Martin
                         Polonaise              S.P.

ELIAS Hermine            26.11.79 Hampiegers    22 rue Eugène Varlin
née SIMON                Allemande              S.P.

FOGIEL Aaron             1897 Brzeziny          13 Bd.Rochechouart
                         polonaise              tailleur

FOGIEL Benjamin          13.1.41 Paris          13 Bd.Rochechouart
                         franc.nat.

FOGIEL Chana-Fraja       16.1.06 Bezeznini      13 Bd.Rochechouart
n4s FEUTZINER            Polonaise              S.P.

FOGIEL Marguerite        11.7.34 Paris          13 Bd.Rochechouart
                         franc.nat.            ecolière

FREUND Markej            6.7.92 Prague          Avenue de la Marne, BIARRITZ
                         Allemande              S.P.
```

Fig. 2.5   List of prisoners in Drancy on staircase 2 who were transported to Auschwitz on 11 November 1942. Note that "Buchinsky" should be "Duschinsky".[64]

German name for Friedrich which he often used, but it is unlikely that he was using a false name of Buchinsky. The substitution of a B for a D is very likely a typographical error, or a modification made in a subsequent correction of the document, as "Buchinsky, Bedrich" in the Drancy deportation list is placed directly below "Bruker, Ida" whose real name was "Druker, Ida".[63] Furthermore, "Buchinsky, Bedrich" and "Bruker, Ida" are placed between letters D and E in the alphabetical Drancy list. The surviving typed records for transportations were in poor condition and it is possible that post-war corrections were needed to attempt to make them legible. Sadly, in the chaos and complications of post-war

Europe, some people became a misspelt name or even a number and were lost to history and forgotten. One aim of this book is to set that right, at least for a small number of people.

Being Jewish people from a country outside of France, who were being prioritised for deportation by the Nazis to "the East," the group from Gablonz was particularly vulnerable. They were not held in Drancy for long and on 11 November 1942, Duschinsky and the three members of the Altmann family were transported in Convoy 45, Train Number Da. 901/38, from Drancy to Auschwitz. The transportation was carried out with military efficiency. A telex from Ernst Heinrichsohn, a German lawyer employed by the German security police, and signed by his superior Heinz Röthke, was sent to SS-Obersturmbannführer Adolf Eichmann stating that the Convoy had left Bourget/Drancy station in the direction of Auschwitz with 745 Jewish adults.[65] The transportation contained 350 males, 391 females and 4 undetermined people. There were 106 children under 16 years of age of whom 63 were under 12 years of age. The convoy passed through Bobigny, Noisy le Sec, Epernay, Châlons sur Marne, Révigny sur Ornain, Bar le Duc, Lérouville, Novéant sur Moselle and Metz in France; then Saarbrücken, Frankfurt am Main, Dresden, Görlitz, Neisse and Cosel in Germany; and then Katowice and finally Auschwitz in Poland.[66] Between 1942 and 1944, there were 76 such transports between Drancy and Auschwitz, which took essentially the same route through these towns in France, Germany and Poland.[66] It is hard to believe that these transports were not noticed by some inhabitants of those towns. Convoy 45 took two days and two nights arriving at Auschwitz on 13 November 1942.

Databases for those deported give the name of Bedrich Buchinsky and also Bedrich Duschinsky with the correct birthdate in both cases.[64,67] The names and birthdates for the three members of the Altmann family were listed with no errors. The conditions for the transportation were appalling with close to 100 people squeezed into each cattle wagon. It was essentially impossible to sit down and there was almost no light, ventilation, food or water. On arrival in Auschwitz/Birkenau, 112 men and 34 women were separated for hard labour. The remaining 599 people were sent to the gas chambers.[63]

The Auschwitz Death Book (the "Sterbebücher"), kept meticulously by the authorities there up to 1943, states explicitly that Friedrich Duschinsky, born on 26 February 1907 in Gablonz, died on 1 December 1942 with his last residence as Bagneaux-sur-Loing.[68] It also recorded that 17-year-old Alexandre Altmann (son of Hilda and Maximilian Altmann), born in Alt-Harzdorf, died on 3 December 1942 with the same last residence of Bagneaux-sur-Loing. There was no mention of the skilful inventor Hilda Altmann or her husband Maximilian in the Sterbebücher. This implies that it is very likely that this couple perished during the terrible journey of Convoy 45 from Drancy to Auschwitz. Only two of the 745 people in the Convoy are known to have survived the war.[61,63,66]

Fritz Duschinsky's parents, Alexander, who was born in Bratislava in 1876, and Eugenie (Jenny), who was born in Liberec in 1887, were also murdered in Auschwitz-Birkenau. It had been the intention of Fritz's brother Erich to find a way to bring his parents to the UK in 1938, but, after the Munich Agreement and occupation of the Sudetenland, this proved impossible.[23] They had moved from Gablonz to nearby Turnov and were transported from there to the internment camp at Theresienstadt on 16 January 1943 and then on to Auschwitz on 6 September 1943.[67] Even though Alexander Duschinsky was a World War I veteran and German nationalist, this made no difference to his tragic fate.

Theresienstadt was a prison camp set up by the Nazis in November 1941 to hold Jews from the Greater German Reich.[22] The camp was based in the town originally called Terezin by the Czechs, and was just 150 km by road from Gablonz. The aim of the Nazis was to use Theresienstadt especially for Jews who were over 60 years old, WWI veterans and prominent people such as artists, musicians and scientists. This is why Theresienstadt was a common destination for several of the more senior scientific refugees and their spouses discussed in this book. The Nazis often used the trick of describing Theresienstadt as a spa town for more senior people and, in this way, they were able to force the confiscation of the assets, houses and belongings of many of the inhabitants of the camp. They also arranged occasional and highly organised visits from outsiders to give the false impression that this "show camp" was habitable (as described later in this book in the section on Paul Eppstein).

The mortality rate for the inhabitants of Theresienstadt was high because of the very poor and meagre food, little health care except that provided by the inhabitants themselves, families separated, almost no heating in winter and highly inadequate sanitary facilities. Over 30,000 prisoners died there during the war.[69] It is remarkable that, nevertheless, some kind of civilised life did continue in Theresienstadt. Lecture programmes were even organised by the prisoners as described in some cases in later chapters of this book. Between October 1942 and October 1944, more than 45,000 people were sent from Theresienstadt to Auschwitz in 25 large transports.[69] There were also several transportations to other infamous death camps, such as Treblinka and Sobibór, and also to notorious ghettos further east, including Lublin and Minsk. Nevertheless, as is described later in this book, a small number of scientific refugees did manage to survive Theresienstadt and were then able to continue their productive careers afterwards.

After the war, the new Assistant Secretary of the SPSL, Ilse Ursell, who replaced Tess Simpson in 1945, worked meticulously to determine the whereabouts of those who had applied for assistance to the AAC or SPSL and had lost contact. She wrote many letters to other agencies assisting refugees and also to the original referees of the applicants whose addresses were known to the Society. On 28 May 1947, she wrote to Professor Karl Weissenberg who was then living in Manchester:

> I am writing to you today to ask whether you have any news of F. Duschinsky, who was your Assistant in Dahlem. Dr. Duschinsky appears to have been a Czech who went to Russia from Belgium but left Leningrad again in 1937 to return to Czechoslovakia. Since then we have not heard anything. If you have any news I should be most interested to have it.[24]

Weissenberg replied that he had no knowledge of Duschinsky's whereabouts but suggested contacting the Czech Refugee Trust Fund. This Trust Fund then proposed that she could contact Erich Duschinsky who was born in Gablonz and was now living at

56 Canfield Gardens, London.[24] Erich Duschinsky, in a letter with the personal heading "Consultant to the glass industry," and a new address of 67 E. Harrington Gardens, replied to the SPSL on 24 January 1948:

> Dr. F. Duschinsky was my brother. Unfortunately, I can only report that he has not returned from Nazi occupied France, and vague information is available that he was taken into one of their extermination camps on account of a denunciation, as late as 1944, in Paris, where he was hiding until then.
>
> I have tried to obtain further particulars, but I have failed, also to trace any remnants of his scientific work, with which he was occupied during the early part of the war.[24]

Ilse Ursell also wrote to Peter Pringsheim who had been the research supervisor of Duschinsky in Berlin and Brussels, and who was now at the University of Chicago. Pringsheim himself was fortunate to be able to leave France during the war. When the German army invaded Belgium in May 1940, he was at once arrested by the Belgium authorities as he was a German citizen. Pringsheim was first sent to the internment camp of St. Cyprien in Southern France and then to another camp at Gurs, which was being run at that time by the French Vichy authorities.[70]

To be given permission to leave a concentration camp such as Gurs, the French authorities required details of a visa to enter a country such as the USA. To obtain such a visa, the American authorities insisted on evidence that work could be started on arrival in the USA. In some cases, evidence of a host in the USA and details of costs to cover the stay of the refugee also had to be provided. In addition, the refugee had to find the substantial expenses needed to cover a berth on a boat to the USA from a suitable port in Europe. There were agencies in the USA which worked hard to assist the refugees in these difficult tasks. As we shall see, the requirements proved to be a very challenging hurdle for several of the scientific refuges discussed in this book, and some were still unable to leave for the USA even if they had accumulated the necessary documents and finances.

Pringsheim's imprisonment in St. Cyprien and Gurs put him into direct or indirect contact with several other scientists discussed in this book who had been interned in France. His brother-in-law was Thomas

Mann, the well-known German author and winner of the Nobel Prize for Literature. Mann himself had moved from Germany to the USA and he enabled Pringsheim to get a formal offer of a position at Berkeley, University of California. At that time, the USA being neutral in the war enabled Pringsheim to be allowed to leave Vichy France and travel to Portugal where he caught a boat in February 1941.[70] His wife Emilia had stayed in Belgium for the duration of the war. Pringsheim responded to Ilse Ursell on 6 February 1948: "I am very much afraid that Dr. Duschinsky has been killed by the Nazis in France."[24] This was the last information that the SPSL had on Fritz Duschinsky and, as far as the author has been aware, there have been no publications since then with more detail on his activities in France until the present book.

On 8 May 2022, a memorial plaque was placed in Bagneaux-sur-Loing for twelve people who were arrested in that town and deported from France.[62] The names on the plaque included Alexandre, Hilda and Maximilien Altmann, and Bedrich Buchinsky (see Figure 2.6). As we have seen, Bedrich Buchinsky was really Friedrich Duschinsky. The name Bedrich Buchinsky is also on the Shoah Wall of Names in Paris.

Duschinsky's last paper, published in *Acta Physicochimica URSS* in 1937, has become particularly influential since the advent of

Fig. 2.6   Memorial in Bagneaux-sur-Loing naming those arrested and deported by the Nazis. The names include Bedrich Buchinsky and Alexandre, Hilda and Maximilien Altmann.

electronic computers.[41,71] The computational method he developed for treating the overlap of molecular vibrations in different electronic states is now known as "Duschinsky Rotation", "Duschinsky Mixing" or the "Duschinsky Effect" and is sometimes also referred to by the name Duschinskii or Duschinski.[72] His computational procedure has enabled spectra to be calculated for many molecules in the visible and ultraviolet regions of the spectrum, and has also had applications in several other areas such as molecular charge transport, electron transfer chemical reactions and photoinduced cooling in polyatomic molecules.[73–75] The title of the paper indicated it was the first of a series, but, sadly, Duschinsky did not have the opportunity to continue this work as he had to leave Leningrad as soon as the work was published. His paper was re-discovered in the 1960s and has now received over 1,000 citations, which is a remarkable record for a publication in a relatively obscure Russian journal of physical chemistry in 1937.[41]

Duschinsky's papers published in 1933 on the theory of fluorometry (often called fluorescence spectroscopy) have also become highly influential. He wrote down the basic equations for extracting information from fluorometry measurements, which nowadays have numerous analytical applications for detecting molecules in very low concentrations with wide-ranging use in biology and medicine.[27–29] Duschinsky died at the young age of 35 and, given his outstanding track record for original research, it is very likely he would have produced more outstanding papers if he had survived the war.

2.2 Emanuel Wasser

There does not appear to be evidence that Fritz Duschinsky and Emanuel Wasser ever met, but their tragic stories have many links.

Emanuel Oskar Wasser (1901–1944) was born in Lvov (also called Lviv and Lemberg) on 9 November 1901 to a Jewish family (he was also called Emmanuel Wasser in France and Emanuijl Waser in Poland). His father was Daniel, born in Lvov in 1875 and his mother Anna (née Schiffmann) was born in Cracow in 1878. Lvov, which was then in Poland and is now in the Ukraine, was situated in the Austro-Hungarian

Fig. 2.7 Emanuel Wasser and his wife Sara Zylberszac in their Belgian Aliens Registration Certificates of 1939.

empire and, accordingly, Wasser was able to take up Austrian citizenship. He studied mathematics and physics at the University of Vienna from 1920. He was awarded his Dr. phil. in 1924 and his supervisor was Professor Felix Ehrenhaft who had taken the prestigious chair previously occupied by Boltzmann in 1920.[4] Wasser then continued this research in Vienna until 1932.

Wasser published several papers, some single author and some with Ehrenhaft, in the *Zeitschrift für Physik* on the electric and magnetic properties of sub-microscopic particles.[76] These works were part of the broad effort from the laboratory of Ehrenhaft on finding particles with charges smaller than that of the electron. In 1926, Wasser and Ehrenhaft published a paper on this topic in the *Philosophical Magazine*, perhaps with an eye on establishing a reputation in English-speaking countries.[77] However, their claim on particle charges was not accepted by leading physicists working in this field such as Robert Millikan. Ehrenhaft was a close personal friend of Albert Einstein who would stay with him on frequent visits to Vienna. However, Einstein did not accept Ehrenhaft's claims on particle charges either. In addition, Ehrenhaft claimed to have observed isolated magnetic poles, but Einstein also disputed this.[78]

Even in 1926, Ehrenhaft was concerned about the future career of his star pupil Emanuel Wasser and wrote on 6 December to Einstein:

> Emanuel Wasser's chances for an assistant appointment are nil, given that he is a Jew and a foreigner, despite the fact that he is an excellent experimentalist. I have applied on Wasser's behalf for a stipend with W. E. Tisdale, assistant director for science with the International Education Board. I request your supporting recommendation.[79(t)]

In the 1920s, Einstein had become the most celebrated scientist and a letter of recommendation from him should have been very influential. However, despite his efforts with Einstein, Ehrenhaft was not successful in obtaining a position for Wasser until 1932 when a significant offer came from Leningrad in Russia. There, Wasser was made Head of the Photoelectric Laboratory of the Physico-Technical Institute of the Urals. In Leningrad, Wasser published several papers on the photoelectric effect in Soviet journals and one paper in English in the *Physical Review* in 1935 on the principle of spectroscopic stability.[80] This fundamental work involved the non-observance of double refraction in thallium atoms due to a strong magnetic field, a result which was claimed to support the principles of quantum mechanics. However, on 26 April 1935, Wasser was shocked to find his position in Leningrad had been terminated without explanation. He returned to Lodz in Poland, which was the hometown of his wife Sara (née Zylberszac). They had married in Lodz in 1929 and she was a key part of Wasser's story.

From Lodz, Wasser wrote to the AAC on 13 May 1935 asking if a position could be found for him.[81] He also wrote to Max Born on 3 May 1935 requesting help with his AAC application and saying that his expulsion had caused panic amongst the foreign scientists in the Leningrad laboratory.[81] As we have seen from the case of Duschinsky, word was getting around about the influence of Born in the UK. Wasser listed

Einstein, Ehrenhaft, Hans Thirring (Vienna) and Jakov Frenkel (Leningrad) as referees. He wrote that his field of expertise involved measurements on sub-microscopic particles, photoelectricity and metal physics, and he should be able to work in an industrial field or in a company where physical and technical problems are dealt with. He described his religion as "Jewish reformed." He also stated that he would be prepared to work in the Far East and South America "if the living conditions there are tolerable and they could offer me sufficient job opportunities."

With the usual process, Tess Simpson then consulted with several UK physicists about Wasser's application. On 7 October 1935, the AAC received a phone call from Patrick Blackett who was then a professor at Birkbeck in the University of London and was to receive the Nobel Prize for Physics in 1948 for his work on cloud chambers. The written record of the phone call described some gossip in somewhat scandalous terms:

> Blackett phoned. Made enquiries in the USSR. Found that a lot of the men in Ioffe's lab in Leningrad were indirectly implicated in the recent political assassination (probably unjustifiably). Wasser's wife who had been making visits to Poland boasted on her return on her relations with Polish officers. This caused suspicion of Wasser and also of Fröhlich, so contracts finished. Thus, Wasser was indirectly the cause of Fröhlich's dismissal.
>
> Blackett would not have him in many areas including in his lab. Thinks we should not take responsibility for his support. Does not know how good he is. Suggest he stays in Poland if his wife has good relations with official circles.[81]

Abram Ioffe who was one of the most highly distinguished Soviet physicists is mentioned in the record of the phone call. He had made broad contributions in many areas including radioactivity, electromagnetism, nuclear and solid-state physics. He had arranged for his brilliant student Pyotr Kapitsa to work with Rutherford in the Cavendish Laboratory in Cambridge. The political assassination referred to in the letter was that of Sergei Kirov in December 1934 who was a close associate of Stalin and was the head of the Soviet Government in Leningrad. This led to purges of many of the elite in the Soviet Union including some physicists.

Herbert Fröhlich (1905–1991), also mentioned in the phone call, was a Jewish physicist born in Germany who had moved to the same laboratory as Wasser in Leningrad after Hitler came to power in 1933. Fröhlich was already making progress in the applications of quantum mechanics to the solid state. Schrödinger had heard about this work and wrote to the AAC on 7 November 1933 from Magdalen College, in his very first week in the UK after moving from Berlin to Oxford, to say he would like to engage Fröhlich as a research assistant.[82] Fröhlich had also impressed the physicists Nevill Mott and Arthur Tyndall at Bristol University who had met him at a conference they organised in 1935. They established, with the assistance of the AAC, a position for Fröhlich in Bristol.[83] In Wasser's letter to Born, he stated that Fröhlich should be able to help make his case in the UK.[81] The AAC also wrote to Paul Dirac, who was visiting Russia in 1935, asking if he could find any information on why Wasser and Fröhlich had been dismissed in Leningrad.

Following the negative tone of the phone call from Blackett, however, the AAC was not inclined to provide assistance to Wasser and a letter on 11 October 1935 from the General Secretary to him stated firmly:

> We have not heard of any opening which may be suitable for you and at the moment can see no other means of assistance … I am afraid that it will be very difficult to find even unpaid facilities in this country partly because there are so many displaced German scholars already here as temporary guests and also because the Heads of laboratories would be unwilling to take implicit responsibility of inviting you to their departments. It might also prove difficult to secure permission from the Ministry for Labour and Home Office for a permit of residence in this country.[81]

Wasser then returned to Austria for a short period, but after the Anschluss on 13 March 1938 he wrote again to the SPSL on 14 April asking for assistance and stating that he had been forced to leave Austria and return to Poland "where I received a permittance of stay only for a short time. What I shall do afterwards I do not know."[81] He also wrote that he would be prepared to work in an "even unpaid facility." The implication of this and other letters is that he had some private means. Tess Simpson quickly replied on 21 April 1938 in the continuing negative tone:

> The possibility of obtaining even unpaid employment in this country is very slight indeed. There are already so many displaced German scholars here and until these are permanently established, the heads of laboratories will not take the responsibility of inviting newcomers. The Home Office and Ministry of Labour are now much more severe in this respect than they were formerly.[81]

After the Anschluss in Austria, the Nazis were moving very quickly against opponents, dissidents and Jews. There were reports that the three Nobel Prize winners in Graz, Schrödinger, Victor Hess and Otto Loewi, had all been jailed.[4] Of these three, Schrödinger was the only one who was not Jewish. This set in motion a frantic set of letters between the SPSL, government officials and several leading British academics including Henry Dale and Patrick Blackett. Following requests from George Gordon, President of Magdalen College, and Frederick Lindemann, Dr. Lee's Professor of Physics, the Foreign Secretary of the UK and Chancellor of the University of Oxford, Lord Halifax, consulted with the German Government at the highest level requesting permission for Schrödinger to return to Oxford where he still held a Fellowship at Magdalen College.[4] The Foreign Minister of Germany Joachim von Ribbentrop replied with a telegram sent to the British Ambassador in Berlin, Nevile Henderson, on 29 April 1938 (translated by the Embassy):

> I have been informed that Professor Schrödinger left Germany in 1933 for political reasons and, after a brief stay in Oxford, settled in Graz. There he proceeded to busy himself as a fanatical opponent of the new Germany and National Socialism. A recently undertaken examination of his case has confirmed the fact that he has up till very lately remained in constant contact with German emigrés living abroad. It is therefore to be feared that permission for him to go to Oxford would merely offer him a further opportunity to take up once more his anti-German activities.[84]

Realising the difficult situation he was in, Schrödinger, who was not Jewish, in September 1938, managed to escape with his wife Anny from Austria to Italy from where he made his way back to Oxford. He then moved late in 1938 to Ghent in Belgium where he met up with Peter Pringsheim. However, unlike Pringsheim, Schrödinger had the good fortune to be able to leave Belgium as soon as the war started and went to a position at the new Institute for Advanced Studies in Dublin.[4]

Both Hess and Loewi were Jewish and were allowed to leave Austria only after paying over their Nobel Prize money.[4] Wasser's supervisor Ehrenhaft was also Jewish and managed to leave Vienna for France in 1938. He briefly then went to England where the SPSL were surprised to find he was living in a hotel in London.[85] He had family in the USA and he spent the war there before returning to his professorship in Vienna afterwards.

The SPSL found it very hard to find placements for the most eminent of the refugee scholars. In his application to the SPSL, Ehrenhaft had named a formidable array of referees including Einstein, Schrödinger, Planck, Nernst, von Laue, de Broglie, Langevin, Richardson, Bragg, Zeeman, Compton, Milliken, Langmuir and Weyl.[85] But this made no difference as it was not possible to find a suitable chair for him. Of all the eminent refugee physicists who came to the UK in the 1930s, chairs were found only for Born and Peierls before 1940. Born had only been given a temporary lectureship at Cambridge and moved to Bangalore in India in 1935. However, the local politics there did not provide a permanent position for him either. A chair in theoretical physics then became available at Edinburgh University in 1936 following the appointment of Charles Darwin, grandson of the famous biologist, to the Mastership of Christ's College, Cambridge.[1] Schrödinger was considered first for this position but, after he had decided to go from Oxford to Graz, it was offered to Born who accepted.[4]

An appropriate position for Schrödinger in Oxford could also not be found. Although the Sedleian Chair of Natural Philosophy had been informally promised to him by Lindemann, the incumbent in that Chair, Professor Augustus Love, an expert on classical wave theory, remained alive and did not retire during the short stay of Schrödinger in Oxford from 1933–1936.[4] Fritz Haber, another highly eminent Nobelist, had also not found it possible to find a suitable position in England when he moved to Cambridge in 1933, and he died early in 1934.[86]

Max Reiss, a less-known physicist who published several papers with Wasser and was a student of Ehrenhaft, did manage to emigrate from Austria. After being unsuccessful in an application to the SPSL, he arranged a position with the Eastman Kodak Company in Rochester, USA, in 1939.[87] There, he worked on lens design and developed the theory of off-axis illumination in the camera image.

Wasser applied everywhere he could and he wrote again to the SPSL on 10 September 1938 asking if assistance could be provided in helping him to obtain a migration certificate to move to Palestine.[81] He said that he had to leave Poland by 30 September 1938 and a forced return to Vienna "could end in a catastrophe to me and my family." The General Secretary David Cleghorn Thomson replied to say that the SPSL unfortunately could not help with a move to Palestine. Thomson, a Scottish politician, had replaced Walter Adams as General Secretary in July 1938. Adams had moved on to become Secretary of the London School of Economics, an institution he would become director of in 1967 before being knighted in 1970.

Following the dreadful scenes of Kristallnacht on 9–10 November 1938, Wasser was not prepared to risk taking his wife Sara and his daughter Elisabeth, who was born in Vienna on 8 May 1932, back to Austria and he tried his luck with Belgium. He had previously met Professor Auguste Piccard of the Free University of Belgium in connection with his high-altitude balloon ascents, one of which famously crashed on a glacier in Austria.[88] Piccard had also made the arrangements for Peter Pringsheim and Fritz Duschinsky to come to Belgium in 1933 and 1934. Wasser arrived with his wife Sara in Brussels on 11 November 1938.[89] The immigration authorities noted that he was an "unpaid researcher" and were not sure about the permission for his entry as he did not have a work permit. However, this was eventually provided four months later (see Figure 2.7). Following the difficulties of being a German citizen in Poland, Wasser had left his five-year old daughter Elisabeth behind in Lodz and his wife Sara wrote to the Belgian authorities on 4 March 1939 asking for permission for her to come to Brussels with a family friend, which was granted.[89]

At the University of Brussels, Wasser collaborated with H. Vogels, a research associate of Peter Pringsheim. However, everything changed on 10 May 1940 when the German forces invaded Belgium. Wasser was at once arrested by the Belgian authorities having had his citizenship transferred from Austrian to German following the Anschluss. Pringsheim subsequently stated in a 1942 paper in *Reviews of Modern Physics* that "further work (on fluorescence of complex ions of thallium, lead and tin) done in collaboration with Messrs H. Vogels and E. Wasser was ready for publication at the time of the German invasion of Belgium. The manuscripts were lost."[90]

Both Wasser and Pringsheim, also a German citizen, were then sent by train to an internment camp at St. Cyprien in the southeast of France, close to the border with Spain.[91] With the capitulation of France in June 1940, the St. Cyprien internment camp initially came under the organisation of the Vichy regime which was established on 10 July 1940. The Belgian government documents state that Wasser's wife Sara and daughter Elisabeth also escaped from Belgium at the same time.[89] On 27 July 1940, Wasser was transferred to the Camp-des-Milles near Marseilles and his wife and daughter had found accommodation in the South of France in Toulouse at 5 Rue Jacques Labatat.[91,92]

In 1940 and early 1941, the internees in the Camp-des-Milles were allowed to visit the US consulate there to attempt to obtain a visa to go to America. To be successful in this, funds were needed to cover the expensive berth on a transatlantic liner for the whole family and evidence was also needed of employment in the USA. Assistance was provided to the Wassers by the US Jewish Transmigration Bureau and substantial funds of $925 were donated in March 1941 by Meyer Freedel for the transatlantic trip of Wasser, his wife and daughter.[92] Freedel was a scrap metal dealer from Pittsburgh who was born in Kisselin, Ukraine, which was quite near Lvov, the birthplace of Wasser. Freedel went on to volunteer to fight for the US army in the war. However, there is no evidence that employment for Wasser in the USA was forthcoming. By the spring of 1941, it was becoming increasingly difficult for refugees in the French camps to immigrate to the USA with the Nazis applying more pressure on the Vichy authorities. For the Wasser family, the emigration attempt did not succeed although the Transmigration Bureau did manage to assist thousands of other refugees.[93]

The next record of Wasser's movements dates from 1943 following a major Gestapo raid which was carried out in Clermont-Ferrand, where the University of Strasbourg had moved in 1939.[94,95] Heinrich Himmler had been annoyed about the University move and the refusal of the University authorities to revert back to Strasbourg after the fall of France in 1940. The reports of resistance activity in the University were also of particular concern to the German leaders. After the capitulation of France, Clermont-Ferrand was situated in the Vichy region but this changed when the Germans occupied the remainder of the country in November 1942 after the Allies landed in North Africa. In June 1943, Alois Brunner was sent to France by Adolf Eichmann to take control of the Drancy internment camp and to arrange for the further roundup and deportation of Jews from France. Brunner also cancelled the opportunity for refugees to leave France if a suitable employment had been arranged elsewhere.[61]

On 25 November 1943, the Gestapo, with the assistance of members of the German Army and Luftwaffe, surrounded the Strasbourg University buildings in Clermont-Ferrand with the main aim of arresting known members of the Resistance, citizens from Alsace-Lorraine, and Jewish students and faculty.[95] After interrogation of over 500 people, 130 were taken prisoner and 110 deported. One of these was "Emmanuel" Wasser and, like Duschinsky, he was sent to the internment camp of Drancy in Paris. The French spelling Emmanuel was used in several records. His birthday was correctly recorded as 9 November 1901.[96] There does not seem to be detailed information on Wasser's activities at the university in Clermont-Ferrand but the record for transportation from Drancy stated that he was a "Lehrer" (a teacher). His address in Clermont-Ferrand was given as 76 Rue des Liondards, which was close to the university quarter.

Being Jewish and a non-French national, Wasser was particularly vulnerable and his deportation to Auschwitz was swift, occurring on 20 January 1944 in Convoy 66.[96] The long train route through France, Germany and Poland was the same as for Convoy 45 taken by Duschinsky. The convoy consisted of 632 males and 515 females, and included 221 children under 18 years of age. On arrival, 236 men and 55 women were taken for forced labour.[66] The 856 remaining, including Emanuel Wasser, were sent to the Birkenau gas chambers. The details of this convoy are

Fig. 2.8   Elisabeth Wasser, photographed on 29 April 1944 by the Police Department in Switzerland.

unusually well documented.[97] The date of Wasser's death was originally recorded by the French Government as 20 January 1944 at Drancy but this was subsequently corrected to 25 January 1944 at Auschwitz.[98] Of those who were on Convoy 66, just 72 survived the war including Suzanne Birnbaum who was to write in detail on the horrific transport, the arrival at Birkenau and many other aspects.[99] The name of "Emmanuel" Wasser is on the Shoah Memorial of Names in Paris together with over 76,000 names of other Jewish people who were deported from France and murdered in World War II.[66]

Wasser's wife Sara managed to remain in France after the arrest of her husband.[100] She was clearly concerned about the safety of her eleven-year-old daughter Elisabeth. Clandestine routes had been set up to take refugee children from France to Switzerland where they were supported by the Red Cross and individuals including Alis Guggenhiem (see also the chapter on Hellmuth Simons). Accordingly, Elisabeth Wasser suddenly appeared in Geneva, without her mother, on 4 April 1944 (see Figure 2.8). In her interview with the Swiss police, she said:

> My mother is still in Grenoble. She wanted me to be safe. She herself feared that she would not be accepted into Switzerland and therefore

stayed in Grenoble. On April 4th I travelled by train via Lyon to Bellegarde. I took the bus from Bellegarde to Sergy. A lady named Roselar took me from Sergy to over the Schweizer Grense. On April 6, 1944, around 13.00, we crossed the Swiss border near Geneva.[100(t)]

Elisabeth also informed the police that "her father was arrested by the Germans in Clermont-Ferrand in November 1943 ... and was deported to Germany."[100] It was clear that neither she nor her mother knew of her father's ultimate fate. When asked if she had any links to Switzerland, she replied that her father had worked with Professor Piccard who was originally a Swiss national. She also mentioned that she and her family had visas to go to America. Elisabeth was then put in foster care with a family in central Zurich (Dr. W. Schmid, Rebbergstrasse 4), which was organised by the Swiss Red Cross. After the end of the war in Europe, Elisabeth was sent back to France on 3 July 1945 with several other refugee children on a train from Geneva via Bellegarde to Lyon (the reverse of the trip she had made to come to Switzerland) where it is presumed she met up again with her mother Sara.[100]

Sara's sister, Anna Zylberszac, was five years younger than her and, like her elder sister, led a very adventurous life. She trained as a doctor in Zurich and married Leon Gecow, also from her home town of Lodz, who was an active member of the Communist Party.[101] Like her husband, Anna became a revolutionary and worked with the anti-Franco units in the Spanish civil war. During World War II, she was captured by the German forces and sent to the Radom ghetto where she worked in a forced labour camp. Freed by partisans in June 1944, she was reunited with her husband who had served in the Soviet Army. She joined a Polish medical unit for the last few months of the war. After the war, she worked as a doctor in a Defence Military Hospital while Leon briefly became Director of the Military Office of the Ministry of Health in Poland. Being German-speaking Jews and having a close connection with the controversial American communists Noel and Hermann Field led to their arrest by the Polish authorities. Anna and her husband were tortured and he died in captivity. After the death of Stalin in 1953, she was released. Anna then worked as a doctor in several other countries including Vietnam and, possibly influenced by her sister Sara, she spent time in France where she died in 1985.[101] The mother of Sara and Anna, Estera Zylberszac (née Rosenberg), had been murdered by the Nazis in Treblinka in 1943 while their father Lajzer had died in 1937.[67]

After the war, Sara applied to the German authorities for compensation following the treatment of herself and her husband Emanuel Wasser. She needed the assistance of the Belgian authorities in this application, who retained details of her original passport.[89] She used her maiden name Sara Zylberszac and died in Paris on 20 November 1991 at the age of 85.[102] Her daughter Elisabeth Wasser always used the surname of her father and died at Saint-Remy-Les-Chevreuse, France on 19 July 2009 aged 77.[102]

In 1948, Ilse Ursell of the SPSL wrote to Pringsheim to ask if he had any information on the fate of Duschinsky. Pringsheim replied on 6 February 1948 that he had heard Duschinsky had been killed by the Nazis. He also stated, "I learned, however, that another scientist Ernest Wasser, who was also with us in France and did not succeed in leaving this country has actually been killed by the Nazis a short time before they fled from this country."[103] Here, Pringsheim mistakenly gives the name of Ernest instead of Emanuel.

A memorial plaque was placed after the war at the University of Strasbourg to remember the 120 people based at their university at Clermont-Ferrand, including "Emmanuel" Wasser, who were arrested and murdered by the Nazis. The day of 25 November 1943 was the largest roundup undertaken by the Nazis in a French University and there are regular ceremonies held to the present day to remember the event. The heading of the plaque states, "L'Université de Strasbourg A Ses Morts Tués à L'Ennemi, Déportés, Fusillés, Assassinés, 1939–1945" (see Figure 2.9).[104]

Wasser's scientific papers have not been as influential as those of Duschinsky. It is interesting, however, to mention that Dirac, the Nobel Prize-winning pioneer in quantum mechanics from Cambridge, pointed out in a 1972 paper entitled "Ehrenhaft, the subelectron and the quark" that there are particles with fractional charges in the case of elementary particles such as quarks.[105] He also mentioned the Ehrenhaft work in his 1948 paper on the Theory of Magnetic Poles.[106] However, Dirac emphasised that the conditions of the experiments carried out by

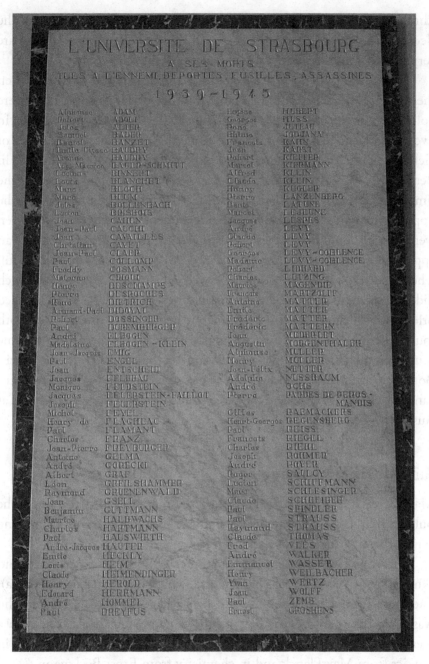

Fig. 2.9   Memorial at the University of Strasbourg listing those from the University who were deported, shot or murdered in World War II. "Emmanuel Wasser" is listed.

Ehrenhaft's research group, including those published by Wasser, on the electric and magnetic properties of small atomic or molecular clusters are very unlikely to have been able to detect quarks.[76,77]

There is no evidence that Duschinsky and Wasser ever met. However, their tragic stories have a remarkable number of similarities. They were both born to Jewish families (Duschinsky in Gablonz and Wasser in Lvov) in places that were far away from the great scientific centres of Germany and Austria where they studied and worked (Duschinsky in Berlin and Wasser in Vienna). Both were recruited to what seemed at the time to be attractive positions at leading institutes in Leningrad with the opportunity to undertake their own independent research. Both, however, were forced out from these positions. Both attempted to get an appointment in the UK through the AAC/SPSL but were not successful. Their research was on the borders between chemistry and physics, which did not seem to link well with laboratories in the UK. Both worked briefly with Pringsheim in Belgium but at different times. They both found themselves as fugitives in France in the war where they became acutely vulnerable after the capitulation of that country to the Nazis. Both were arrested in roundups by the Gestapo and sent to the internment camp in Drancy from where they were transported to the gas chambers at Auschwitz-Birkenau.

2.3 Marie Wreschner

Marie Wreschner (1887–1941) worked and remained in Berlin where she published a far-sighted patent describing an original method for treating cancer through the encapsulation of drugs.

Wreschner was born on 20 September 1887 in Inowrazlaw, Poland (known from 1904–1919 as Hohensalza, which was then part of Prussia). She came from a Jewish family and her father Jakob was a banker who was later based in Berlin. Her mother's name was Paula. She was a German citizen and studied at the University of Berlin taking courses in physics from Max Planck and in chemistry from Ernst Beckmann.

Wreschner's thesis for her Dr. phil. was entitled "About rotation reversal and anomalous rotational dispersion" and was published in 1918.

Fig. 2.10 Group at the Kaiser-Wilhelm-Institut für Physikalische Chemie und Elektrochemie, Berlin-Dahlem, in the 1920s. Marie Wreschner is in the second row, second from the right.

Her supervisor was Arthur Wehnelt who invented the oxide cathode widely used in vacuum tubes. Between 1920 and 1933, she had a research assistantship with Herbert Freundlich at the Kaiser-Wilhelm-Institut for Physical Chemistry and Electrochemistry in Berlin (see Figure 2.10).[107–109] There, she interacted with some of the greatest physical chemists including Fritz Haber and Michael Polanyi. Freundlich's main research interest was in colloids and he published a paper with Wreschner in 1921 on the electrocapillarity of coloured solutions.[110]

In 1925, Wreschner and Laurence Loeb filed a remarkable patent which could be described as a very early example of nanotechnology with a medical application for treating cancer with encapsulated radioactive materials (Figure 2.11). The patent stated:

> Our invention refers to a new composition of matter capable of emitting β-rays and its particular object is to provide a preparation which is insoluble in body liquids, i.e. liquids forming part of or secreted by the

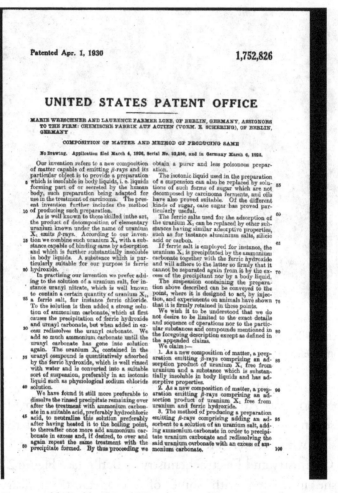

Fig. 2.11   Patent published by Marie Wreschner and Laurence Loeb in 1930 describing a new treatment for cancer.[111]

human body, such preparation being adapted for use in the treatment of carcinoma.[111]

The special feature of their invention was the adsorption of electron-emitting uranyl compounds in a suspension of ferric hydroxide. This approach of encapsulation of drugs in attachments or cages of benign molecules has become common in modern medical applications, and the patent of Wreschner and Loeb was perhaps nearly 100 years ahead of its time. It is notable that both Wreschner and Hilda Altmann, the associate of Fritz Duschinsky, took the innovative approach of publishing via patents and not the more academic method of articles in scientific journals.

Wreschner also undertook another activity of translating an organic chemistry textbook of Robert Robinson from English to German. Robinson was the Waynflete Professor of Chemistry at Magdalen College, Oxford, and was the most influential British organic chemist of that time. He went on to win the Nobel Prize for Chemistry in 1947.[112]

It is clear that Wreschner was allowed to undertake her own independent scientific activities, and photographs of that period show her in a happy laboratory (see Figure 2.10). However, as was the case with so many others, everything changed when Hitler came to power in 1933. Frederick Donnan, Professor of Chemistry at University College London (UCL) who was very much involved with the activities of the AAC, visited Freundlich in Germany in May 1933 with William Rintoul of the Imperial Chemical Industries (ICI) and offered a position at UCL to Freundlich. ICI were aware that Freundlich's work had the potential for lucrative industrial applications to pastes and adhesives. Freundlich asked if he could bring one of his research assistants to UCL, Dr. Karl Söllner, who had already published as many as 28 papers, several with Freundlich. This was readily agreed upon but, unfortunately, there was no such position arranged for Wreschner.[113] Both Freundlich and Söllner eventually moved to the USA in the late 1930s.

Nearly all the eminent professors at the Kaiser-Wilhelm-Institut for Physical Chemistry and Electrochemistry in Berlin in the early 1930s were Jewish and had to leave Germany. This included the Director and Nobel Laureate Fritz Haber (who initially went to Cambridge), Michael Polanyi (who took a chair in Manchester) and Freundlich. The Nobel Laureate James Franck, who had been a Professor of Atomic Physics at the Kaiser-Wilhelm-Institut in the 1920s, and his replacement Rudolf Ladenburg also both immigrated to the USA in 1935 and 1932, respectively. Following the forced emigration and death of Haber in January 1934, a Nazi sympathiser Peter Adolf Thiessen was eventually appointed in his place and the research became directed to problems of military interest.[108]

After 1933, Wreschner was in a very difficult position and was partly dependent on the financial support from her mother who she cared for at Claudiusstrasse 3 in Berlin. Her versatile skills were put to good use by Emil Abderhalden who commissioned her to write articles for his multi-volume *Handbuch der biologischen Arbeitsmethoden* (*Handbook of Biological Working Methods*).[114] Abderhalden was the President of the German National Academy of Sciences, Leopoldina, and had invented a controversial pregnancy test.

By 1938, things were getting very threatening for Wreschner and she wrote to the SPSL for assistance, her application being received on 28 November 1938.[115] She gave her referees as Freundlich, Abderhalden and Bruno Lange, the inventor of the selenium photocell. In her application, Wreschner stated that her medical invention using β-rays was having some therapeutic success but the work was stopped in 1933. She also emphasised the potential of her laboratory report of 1938 on "Photoelectric Cells and Their Applications." This was written in association with Bruno Lange whose company, now called Hach Lange, has grown significantly and is still involved in photocells and their wide-ranging applications. In her report, Wreschner was again scientifically many years ahead of her time. Wreschner stated, "I should be able to do useful work with photoelements and the newest photoelectric apparatus … I should be very glad to do experimental work or literary scientific work of any kind, and I should be able to work as an assistant in scientific libraries or as a teacher of science."[115]

Although Wreschner had written several research reports in Berlin, she had few publications in the formal scientific literature. This put her at a disadvantage in seeking assistance from the SPSL to find her a suitable position in the UK, even though her patent and reports were right at the cutting edge with huge potential applications. The SPSL had also received very few applications from women. In addition, her former research supervisor Freundlich had just moved in 1938 from London to Minneapolis. This diminished his influence in the UK despite the fact he was to be elected in 1939 to the rare honour of Foreign Member of the Royal Society.[116] His very strong reference for Wreschner, written on 1 July 1933, stated:

Dr. Marie Wreschner worked from 1920 until now at the Kaiser-Wilhelm-Institut for Physical Chemistry and Electrochemistry in the Department of Colloid Chemistry of which I have been the Director. At first, she was engaged in purely scientific research on problems of capillary chemistry, amongst others on the adsorption of radioactive elements on charcoal. This research led on to an investigation, which she carried out in collaboration with Dr. Farmer Loeb, using suspensions of iron oxide particles with adsorbed Uranium for the treatment of cancer. By careful and untiring work Dr. Wreschner succeeded in producing suspensions of this kind which in animal experiments were markedly successful in destroying malignant tumours, though perfectly harmless in other respects. Although she tried with great energy and patience to induce medical circles to test this method on a larger scale on man this has not so far been done, probably owing to the difficult times just now. In the last few years, Dr. Wreschner was mainly engaged in literary activities, both translations and monographs, a series of the latter has been published in Abderhalden's, "Handbuch der biologischen Arbeitmethoden".

Wreschner carries out every investigation with great industry, patient perseverance and clear understanding. She is absolutely reliable and personally very pleasant to deal with. I therefore would like to recommend her most warmly.[115(t)]

Freundlich's recent colleague at UCL, Donnan, wrote an additional letter of support to the SPSL on 2 February 1939 stating, "There is no doubt that this lady has done some excellent scientific work and is very highly qualified in science."[115] On the same date, the SPSL received another letter from Wreschner describing her translation of the organic chemistry text book of Robinson and suggesting that he could provide a reference on her behalf.

By 1939, the SPSL had been flooded with applications from many countries. Tess Simpson wrote to Donnan on 6 February, "We are in direct contact with Dr. Wreschner but until now have not been able to find anything for her. We suggested that she approach the International Federation of University Women, but we have not heard whether this has been done."[115] Tess Simpson also wrote to the Jewish Professional

Committee in London asking if they could find assistance. She said on 28 February 1939, "The various institutions and laboratories are now getting so full that I think they prefer to give what facilities they can to younger people who have a chance of getting off their hands within a limited period."[115] In a letter on 9 March 1939, to the Emergency Sub-Committee for Refugees, British Federation of University Women, Tess Simpson also stated, "It seems to me that the chief difficulty in Dr. Wreschner's case is her age."[115] By 1939, Wreschner was aged 51 and, although this was not as old as some of the distinguished applicants to the SPSL discussed in this book, her age was still counting strongly against her. There was some support for her from elsewhere in the UK and a leading refugee chemist Professor Friedrich Paneth wrote on 4 March 1939 to the SPSL from the University of Durham, "I do not know Dr. Wreschner personally but she has been recommended by Prof. Arthur Rosenheim, Berlin ... Besides experimental investigations she has devoted much time to the compilation of comprehensive scientific articles which have been received very favourably."[115]

Wreschner then wrote to the SPSL again on 20 February 1939 emphasising once more her translation of Robinson's book and her highly varied and applied work on the treatment for cancer, photoelectric cells and electric measuring. She also stated, "As you suggested domestic work, I wish to tell you that I have done all the domestic work for my mother and myself since 1933, but I did my literary scientific work as well. I am able to do domestic work (most scientists are) and I should not mind doing it in England as a beginning, but I should be very sorry indeed to stop scientific work for ever."[115] None of the male applicants to the AAC/SPSL had domestic work suggested to them. This suggests that not only was her age a disadvantage to Wreschner in her application to the SPSL but also the expectations of her gender. In addition, having to look after her mother, which was mentioned on her application form, made the opportunity for Wreschner to emigrate much more complicated.

The British Federation of University Women also took up Wreschner's case and wrote many letters to try to assist her.[117] From Oxford, the Federation were told on 7 March 1939, "With Prof. Robinson her work is not quite on his lines and anyhow he has no vacancies in his department. However, he has very kindly agreed to appeal on her behalf to Prof. Hinshelwood and Prof. Lindemann, both of Oxford."[117] Robinson's Nobel Prize was awarded for his synthesis of organic molecules of biological interest and he clearly felt Wreschner's interests were more on the physical chemistry side despite the fact that she had translated his book on the Electron Theory of Organic Chemical Reactions.[112] He was not averse to having refugees in his research group and was to shortly supervise two of the first students at Oxford from China, T. E. Loo and Rayson Huang, who escaped from the Japanese army when they invaded Hong Kong in December, 1941, and trekked right across Asia to come to the UK.[118] Hinshelwood also was to win the Nobel Prize for Chemistry, in 1956, for his work on unimolecular chemical reactions. He wrote on 16 March 1939 to his Oxford colleague Robinson, "I have looked over the papers of Dr. Marie Wreschner and I must say it seems to me a more difficult case than usual. She is really a physicist and off my line almost as much as off yours ... I should have thought Freundlich could perhaps introduce her to the London people."[117]

Frederick Lindemann, the Dr. Lee's Professor of Physics at Oxford, had done much to assist refugee physicists to come to Oxford including Schrödinger and Francis Simon with several of his associates from Breslau.[4] It seems that there is no record of a response from Lindemann about Wreschner. It is clear that Wreschner's interdisciplinary work fell between the traditional subject boundaries, not being considered a chemist or a physicist. On 9 August 1939, a more promising letter was sent to the SPSL from the medical side. Dorothy Parsons of the Royal Cancer Hospital in London was working on the effect of X-rays on blood and lymphoid tissues. She had heard about Wreschner's innovative work on cancer therapeutics. She stated, "She is so fully qualified in physics and chemistry and has had such a wide experience in the application of her knowledge to cancer that I feel an opportunity should be given to her to continue her good work. If I had not guaranteed I am supporting a Czech refugee just at this moment I could offer some assistance."[115] Tess Simpson responded asking whether there might be other hospitals where Wreschner could work. This was, however, just one month before the start of the war and the opportunity for the SPSL and other agencies in the UK to help Wreschner had passed.

Early in 1939, Wreschner had also written to Max Bergmann, a biochemist who had undertaken research at the University of Berlin in the early 1920s and may have met Wreschner then.[119] He had left Germany in 1934 to take up a post at the Rockefeller Institute for Medical Research in New York. Bergmann did also consult about Wreschner with Freundlich who now was in Minneapolis. However, there was no positive outcome. There is no record of an application from Wreschner to the Emergency Committee in New York.[12]

Wreschner's situation in Berlin now became acute. There is one report that she and her mother were robbed and made to undertake forced labour.[107] Her mother then died on 18 September 1940.[107] Wreschner continued to live at Claudiusstrasse 3 in Tiergarten which was a neighbourhood of Jewish people in Berlin. By the autumn of 1941, over 300,000 people classified as Jews remained in Greater Germany. Up until then, Adolf Hitler had been reluctant to order the deportation of Jews from the German Reich because of an expected resistance from the German population. However, with the start of war against Russia on 22 June 1941, the atmosphere at once became more dangerous and Hitler gave the order in September 1941 for the deportation of all Jews still in the Greater German Reich and its Protectorates. There were over 600 such transportations carried out by train over the next three years.[120] Accordingly, on 7 November 1941, all those living at Claudiusstrasse 3 received a notice from the Jüdische Kultusvereinigung zu Berlin on their transportation N. 4218 from Berlin by train:

> This evacuation is in prospect by order of the authorities for 12 Nov. 1941. Apart from you, this also affects your unmarried family members, in so far as they are in Berlin. We sincerely ask you not to let the severity of this measure tempt you to take ill-considered steps. If you keep calm and prudent, you will make the path easier for yourself and your companions. The following must be observed for the implementation of the measures:
>
>   You can take about 50 kg of luggage. Leave behind liquids, bring items that are really necessary. Warm things are particularly important

such as warm laundry, sturdy shoes, coat. Large luggage should be loaded into luggage trolleys. Hand luggage should include only those items which each person is able to carry with them without taking up more space in the compartment than they are entitled to. It has been ordered that the following items are to be packed in the large luggage: cutlery with the exception of spoons, small tools, hammer, pliers, nails, scissors, pocket knife, shaving kit, darning thread, sewing thread, repair kit, flashlights, tallow candles, matches, medicines except for poisons.[121(t)]

This transport, with similar ones being organised from several other centres in Germany, was to a new ghetto in Minsk, Belarus, which had recently been occupied by German forces in the war against Russia. The list of those transported on "Welle V" which occurred on 14 November, did not survive and nor did most files from the Minsk Ghetto. However, such transportations involved a declaration of the confiscation of assets of those transported, which enables some information to be obtained.[121] Over 1,000 people were on the transport, the oldest being 84 and the youngest 18 months. The Minsk Ghetto, being in the active war zone, was in an appalling state and only four of those transported survived the war. 61% of those transported were women, reflecting the large number of Jewish men who had died fighting for Germany in World War I. Included in the names of those transported were Alice Happ, her sisters Gertrude and Hedwig, Meta Brav, Machol Itzig and Charlotte Lazarus who were all from the same address of Claudiusstrasse 3 as Wreschner.[121] However, the name of Marie Wreschner was not on the deportation list. This is because she took her own life to avoid the transport.

On the Register Office card, her death was recorded as 17 November 1941, three days after the transport left.[122] A more detailed Death Register of the Berlin Registry Offices stated that she was found dead in her apartment on 18 November at 1.50 pm.[123] Another publication stated that she committed suicide by gas.[109] Word got round about the deportations and the death camps, and a large number of Jewish people in Germany took their own life in the same way as Marie Wreschner to avoid transportation.[120] After the war, Max Born constructed a poignant list of 35 relatives, friends and colleagues who were victims of the Nazis, with as many as 14 of these recorded by him as having committed suicide.[124]

On 12 April 1947, Ilse Ursell from the SPSL wrote to Dr. Erna Hollitscher at the British Federation of University Women who had tried to help Marie Wreschner back in 1939. Ursell said:

> I am sending a note concerning Dr. Marie Wreschner formerly of the Kaiser-Wilhelm-Institut for Physical Chemistry in Berlin. The difficulty in placing her lay mainly with her age. We have made enquiries on her through the International Red Cross but have so far received no reply. If you have any information in your records, I propose to write to a number of physicists who may have known her. Professor Lise Meitner for instance may have some news on her.[115]

Dr. Hollitscher replied that they had no information but Elisabeth Schiemann could be asked. An anti-Nazi, Schiemann had still managed to maintain a research position in genetics at the University of Berlin during the war. Schiemann replied that she had consulted with Lise Meitner who said she had known Wreschner but had no information on her fate.[115] That is the last letter on Wreschner in the SPSL file.

In 2015, it was announced by the Marzahn-Hellersdorf District Assembly of Berlin that five streets in a new Clean-Tech Business park would be named after pioneering women scientists.[125] The names were Clara Immerwahr-Haber, Marie Wreschner, Cecilie Fröhlich, Gertrud Kornfeld and Marga Faulstich. Clara Immerwahr-Haber undertook catalysis research and was the wife of Fritz Haber. She had shot herself after Haber directed the first use of chemical weapons in World War I. Cecilie Fröhlich was a mathematician and engineer. She also was Jewish and initially had an industrial appointment in Berlin. She escaped to Belgium in 1937 and, after hiding in France as the Nazis invaded, managed to travel to the USA.[126] There, she had a distinguished career as a professor at the City College of New York and became the first female chair of a University Department in the USA.

The comparison between Marie Wreschner and Gertrud Kornfeld is illuminating. Kornfeld was born in Prague on 25 July 1891 to a Jewish family. She was four years younger than Wreschner. She studied chemistry at the German University in Prague and, from 1919–1923, became an assistant to Max Bodenstein in the Technische Hochschule in Hannover researching on chemical kinetics. When Bodenstein took the chair of Nernst at the Friedrich-Wilhelms University of Berlin in 1923, Kornfeld followed and became a Privatdozent, which allowed her to receive fees from students for her lectures. She was the only female chemist in such a position appointed in the Weimar Republic.[127] She also became a Professor of Chemistry at the University of Berlin and taught numerous courses. In this period, Kornfeld published several single-author papers on reaction kinetics in chemistry journals. She also published in the more popular journal *Naturwissenschaften*, claiming an effect of a magnetic field on the rates of chemical reactions, but that paper was subsequently withdrawn.

Following the new laws of 1933 introduced by the Nazis, Kornfeld was fired from her position in Berlin. With the help of the AAC, she managed to obtain a temporary position in the Department of Physics at University College Nottingham.[128] Her referees included Bodenstein, Franck and Friedrich Paschen. She was also recommended to Nottingham by Freundlich, Wreschner's research supervisor who had been impressed by Kornfeld's research output of over 20 papers. At Nottingham, Kornfeld published a paper in the *Transactions of the Faraday Society* on the photochemical decomposition of sulphur dioxide.[129] She wrote to the SPSL, however, to say she was unhappy with her opportunity to undertake research at Nottingham and she had expected her appointment to be a permanent one. She was only funded for one year and her position was not renewed. It was then sometimes possible for a professor in the UK to find the funds and facilities for a postdoctoral assistant in a temporary research assistantship but a major permanent appointment was very difficult to obtain. Having had the status of a professor in Berlin, Kornfeld hoped for a similar position in the UK. The AAC consulted about Kornfeld with Francis Simon at Oxford, a physicist and refugee from Breslau, and he was quoted as saying, "She would be better in some women's college in America rather than doing research."[128] This indicates that, as was the case with Wreschner, there was a general antipathy towards supporting female scientific refugees for research positions.

Kornfeld's spectroscopic expertise had been noted by Herbert Dingle in the Department of Astrophysics at Imperial College London. She moved there in 1934 and published on the ultraviolet emission of sulphur dioxide which had some astrophysical interest.[130] However, her temporary appointment was uncertain and she decided to do the same as Schrödinger by leaving the UK and moving to Austria.[4] Her position at the University of Vienna was associated with the physical chemist Herman Mark with a grant funded by the American Association of University Women (AAUW). This grant only lasted one year and by May 1936 she was writing again to the UK (now to the SPSL) stating that the economic and political situation in Austria was preventing her from obtaining a position and asking if any help could be provided to assist her in finding a post in the USA.[128]

The AAC consulted with the AAUW who enabled Kornfeld to get a temporary visa to visit the USA. She did this in 1937 and then managed to obtain a position at Eastman Kodak in Rochester, New York, where her experience in photochemistry was found to be an asset. There, she published several papers on chemical kinetics linked to photography, including a single-author work in the prestigious *Journal of Chemical Physics* on the limits of spectral sensing on infrared photography.[131] She was the first female scientist appointed to a permanent leadership position at the Kodak Research Laboratory and directed her own research group. However, Gertrud Kornfeld still suffered discrimination against her gender. When a security guard at Kodak informed her, "No ladies after working hours," she replied, "I am not a lady, I am a scientist."[132] She latterly published several papers in the *Journal of the Optical Society of America* on photographic latent images. In 1948, Kornfeld was elected to the New York Academy of Sciences. She died on 4 July 1955 at the age of 63.[127]

Kornfeld's research field in Germany was quite close to that of Wreschner but she benefitted from applying to the AAC much earlier (in 1933) when at least some temporary positions were available in the UK. She published several papers in international journals and was also proactive in making links with the USA, which proved crucial for her subsequent career. There is no doubt, however, that the highly talented Marie Wreschner would have continued her original scientific work, like Gertrud Kornfeld and Cecilie Fröhlich, if she had been able to leave Germany.

2.4 Alfred Byk

Alfred Byk (1878–1942) was an unusual scientist in that he started his research career as an experimental organic chemist with the Nobel Laureate Emil Fischer but then moved over to the different field of mathematical physics working with Max Planck. A professor and contemporary of Einstein in Berlin, he was unable to leave Germany.

Byk was born on 4 March 1878 to a distinguished Jewish family. His father was Dr. Heinrich Byk (1845–1923) who set up a successful company and factory in 1873 making chemical and pharmaceutical products, including sleeping pills. His uncle Eugen Bamberger (1857–1932) was a well-known organic chemistry professor at the ETH in Zurich. His cousin Suse Byk (1884–1943) became a leading portrait

Fig. 2.12   Alfred Byk (~1910).

photographer in Berlin in the 1920s and her work included some of the most famous photographs of Albert Einstein.

Alfred Byk studied chemistry, physics and mathematics at the universities of Berlin and Freiburg. His first research was in organic chemistry with Emil Fischer, the second Nobel Prize Winner for Chemistry in 1902, and he worked on the synthesis of pyrimidine.[133] An interest in the stereochemistry of optically active compounds then drew Byk in the direction of physical chemistry. He worked on the conduction of gases with Nernst in Göttingen and then on photochemistry in Charlottenburg, Berlin. Byk was made a Privatdozent in physical chemistry for one of the first theses on the absorption of circularly polarised light by racemic compounds.[134] In 1906, he worked again with Nernst, who had now moved to the University of Berlin. His Habilitation with a thesis on "The Equations of State and their Relationships for Thermodynamics" was granted on 17 February 1906. In due course, he became an assistant in Berlin to Max Planck from 1909–1912. His predecessor in this prestigious role was Max von Laue and his successor was Lise Meitner. The Habilitation gave a degree denoted "Dr. Habil." and was a necessary qualification to become a university professor. Byk's wife Hedwig (née Fraenkel) also had a doctorate in chemistry from Freiburg, which was not then possible for women at the University of Berlin. They were married in 1911 and they had two daughters Hilde and Marianne. Although from a Jewish family, Hedwig was baptised and brought up her daughters in the Lutheran religion.

Byk was given the title of Prussian Professor in 1910 at the young age of 32 (see Figure 2.12). His research became more involved with atomic structure and quantum theory, although this was before the papers of Heisenberg and Schrödinger that revolutionised those subjects. On 23 October 1914, together with over 3,000 other academics, including Max Planck and Fritz Haber, Byk signed the Declaration of the Professors of the University and Technical Colleges of the German Empire, which supported the Prussian and German scientific efforts of the First World War.[135] Albert Einstein refused to sign this declaration and signed one supporting pacifism instead. Byk was awarded a medal in World War I for his inventions of inks used for secret messages.[136]

On 13 January 1921, Albert Einstein, by now well established as a professor in Berlin and shortly to win the Nobel Prize for Physics, wrote to Max Born, "Have a look at the short paper by Byk on the law of corresponding states and quanta in the *Phys. Zeitschrift* — it is a nice piece of work."[137(t)] Born replied on 12 February 1921, "I have read Byk's paper and discussed it with Stern. We were not, however, particularly enthusiastic about it; after all, it is only the beginning of a beginning of a theory."[137(t)] Born may have recalled these critical comments when he was asked some 13 years later to give his view on Byk's scientific reputation by the AAC. Following the war, Byk was appointed, it seems with Einstein's support, to a non-official associate professorship at the University of Berlin in 1921 and at the Charlottenburg Polytechnic Institute (now the Technical University of Berlin) in 1922. He obtained a salaried professorship at the University of Berlin in 1923 "for lectures on the boundaries of physics and chemistry." He also became the technical advisor to the General Electric Company of Berlin. He was therefore an established professor in Berlin with a good income. This comfortable position may have contributed to a reduction in his subsequent research output. In 1926, he wrote a chapter on "Theory of Molar Thermodynamic State Variables" for the widely used *Handbook of Physics*, published by Springer. However, he did not redirect his research significantly into the new quantum mechanics and this limited his ability to undertake competitive research in atomic physics. With his chemical background, he might have had the opportunity to apply Schrödinger's new wave mechanics to molecular bonding, as had been done by Heitler and London, but he did not take this opportunity.[138] After 1925, Byk only published two research papers but he did author several patents on the long-distance conduction of electricity, indicating the shift in his interests to applied projects.

Following the imposition of the new laws for the Civil Service in Nazi Germany, Byk was dismissed from his professorship on 30 September 1933 due to his Jewish birth. Anticipating his dismissal, he wrote in English on 8 September 1933 to the Emergency Committee in Aid of Displaced Foreign Scholars based in New York City:

Advised by the Zentralstelle für Jüdische Wirtschaftshilfe at Berlin, especially by Dr. Epstein and Dr. Karpf, I beg to apply for your assistance.

I am a Jew being on leave from my profession as a salaried professor at the University of Berlin and should be very glad to get employment in the USA. I am a mathematical physicist, a special branch which I reached after being first engaged in physical chemistry.

I am informed, however, that a permanent employment at an American university is not likely to be got at this time. For this reason, I may emphasize that I had since 1922 permanent employment as a consulting engineer in the Allgemeine Elektricitäts-Gesellschaft of Berlin (the General Electric Company), one of the two great German electro-technical firms. This position I lost but recently. I was engaged there in mathematical problems of telephony. Therefore, I should be fit for an appointment at an adequate American establishment. The most prominent American firm in the domain of telephony being the Western Electric Company, 195 Broadway, New York City ... But as I do not know if I have chances especially with this company I take into consideration likewise any other American establishment great enough to employ a mathematical physicist for special problems. What I beg you to do for me in this respect is to bring me into contact with the authorities of suitable firms ... I should be very obliged to you if you would give me your opinion about the topic.[139]

Here, Byk mentions being given advice in Berlin by Dr. (Paul) Eppstein (misspelt Epstein) who is discussed later in this book. On 19 September 1933, Byk received a response from Edward Murrow the Assistant Secretary: "Unfortunately, it is a little difficult to be very optimistic as to the results of our efforts in your behalf. As you know, conditions are bad in the States just now, and our own funds are practically exhausted."[139] Byk's papers were sent, along with those of several other applicants, to Bernard Flexner of New York who was known to be advising institutions on their appointments in mathematical physics. It seems that the US Emergency Committee, however, did not have the backing and resources to deal with applications in the detailed and personal way that was being applied by the AAC in the UK. There does not appear to be evidence either that the Emergency Committee was able to approach US companies such as Western Electric about an industrial appointment for Byk.

Byk then wrote to the AAC on 10 December 1934.[140] The referees he listed were among the most internationally famous German physicists and chemists, including Planck, von Laue, Nernst and Bodenstein, and also international leaders such as Tolman (California), Keesom (the Netherlands) and Levi-Civita (Rome). Somewhat surprisingly, his former colleague Einstein, who had left Berlin in 1932 and with whom he had cordial relations, was not listed as a referee whereas several other academics discussed in this book gave Einstein as a referee even if they were not so closely associated with him. As was the case with his application to the US Emergency Committee, Byk stated, "Though I would prefer very much an academic activity, the type of industrial position which would be suitable is that of a technical advisor for mathematical–physical problems of a great electrotechnical firm." He also stated that his religion was "Jewish Reformed" and "for the beginning I can go alone, but in the long run I am obliged to take after me three adult persons, my wife and two daughters." In addition to organic synthesis, Byk listed over 30 of his papers on thermodynamics, the old quantum theory before 1926 and reports on the transmission of electricity. As we have already seen for Duschinsky, Wasser and Wreschner, with his highly interdisciplinary papers on the borders between physics and chemistry, Byk was a challenging case to appoint in the UK. In addition, with very few recent papers and, essentially, none on the new quantum mechanics, he did not provide a compelling case to the leading mathematical physicists. He listed as many as nine patents published in Germany related to electricity transmission, but this would not have been of interest to the British professors working on fundamental physics.

In 1934, Byk was aged 56 and, similar to Wreschner, his age was against him. As was often the case with the AAC for applications from German physicists, Max Born was asked for his opinion. In his usual frank style, he replied on 21 April 1934: "Byk is an older man, very good mathematician, but not gifted with a great physical intuition. I don't know whether he would accept a call to Russia."[140] This blunt opinion meant that the AAC would not give Byk's application priority. The highly influential Chaim Weizmann contacted the Board of Deputies of British Jews to state that Byk had applied for a post in Jerusalem and asking for more information. This was also communicated to the AAC.[140]

Fig. 2.13   Alfred Byk (~1938).

The photograph in Figure 2.13 of Alfred Byk was taken around 1938 and he is clearly looking more apprehensive than in the photograph of the confident younger professor in Figure 2.12. His two daughters Hilde and Marianne were now in their 20s. With support from their mother, but not, at least initially, from their father, they planned to emigrate.[136] They managed to obtain visas to immigrate to Australia. They had been assisted in this by Captain Frank Foley from the British Embassy who subsequently became a British Hero of the Holocaust for helping thousands of Jewish people escape from Germany. They had known Foley because the Byk family had an apartment in Berlin which was let to the British Embassy.[136]

The daughters travelled from Berlin to England, leaving on the morning of the infamous Kristallnacht, and called at the SPSL in London on 18 November 1938. They said that they were "anxious to get father out of Germany." However, the note of the meeting from the SPSL stated "there is nothing very hopeful to say."[140] On the next day, the Byk daughters embarked on the French ship *Normandie* from Southampton to New York, with the stated intention of travelling on from there to

Australia. By travelling first to New York, they may have been hoping to find a way to stay permanently in the USA.[136] While in America, they contacted Albert Einstein, who had been a close colleague of their father in Berlin. He sent them a letter of support on 2 December 1938 from Princeton stating in his typical elegant and thoughtful style, "I should be most happy if the two courageous young women would find opportunities for work in their new home. Professor Byk is a prominent scientist in the field of physics. Every organisation or individual who is helping his daughters is helping also indirectly my dear friend and colleague."[141] The daughters then travelled on to Australia. There, they did not mix in Jewish or German circles, which was deemed suspicious by the Australian police who watched their movements.[136]

Byk made enquiries to see if he could take up a position in Shanghai, but the Nazis did not allow him to move to a position in China. Byk was taken in for questioning by the Gestapo in Berlin towards the end of 1939, possibly because he was managing the property of his family. Byk's wife Hedwig was from a wealthy family which enabled her husband to have some funds to continue to live in Berlin even after he had been dismissed from both his university and industrial appointments. She shared the ownership of a fine house at 52 Kurfürstenstrasse, Berlin, with her sister Anna Graeffner. Anna had taken her own life, officially recorded as due to "drug poisoning," on 3 December 1938 soon after her husband had been arrested on Kristallnacht.[142] Following anti-Semitic court rulings on the "Aryanisation" of culture and property, Hedwig Byk was ordered to sell the house to the Reichsverband organisation.[143] She refused to do this and the pressure on her is a possible reason for her suicide. Her death certificate recorded the date as 12 December 1939 and her death as being due to her "falling from a window."[144] She died in a nearby sanatorium at 16 Achenbachstrasse. At this time, her husband was still under arrest and this must have added to her despair.[136] Alfred Byk and Ernst Graeffner, the husband of Anna, then became the legal executors of 52 Kurfürstenstrasse on behalf of Hilde and Marianne and were forced to sell the property.[136]

The lawyer and journalist Dr. Curt Rosenberg (1876–1966) was a cousin of Alfred Byk and had been one of his closest friends since boyhood. In an unpublished recollection "Aus Der Emigration," Rosenberg wrote of the last time he saw Alfred on 18 August 1939, which was just before Rosenberg left Germany:

> The mood was a bit depressed, but he wanted to remain optimistic because he [*i.e. Alfred*] had to and wanted to stay in Germany in order to pursue his extensive financial interests. He actually sacrificed himself to those interests and never got out. His wife took her own life soon after the outbreak of war. I don't know what became of him. When we left the Byks at about 11 o'clock, I was determined to travel the next day [*to England*].[141]

In the 1940s, the only approved way to send or receive messages between countries at war was through the International Red Cross. Messages were sent via Geneva in neutral Switzerland and only 24 words were allowed in, at most, one message each month. The subject matter and language allowed was highly censored. The messages took several months to arrive at their destinations and the tension of waiting for these messages would have been extraordinary, especially when the messages stopped arriving. On 30 May 1942, Byk sent the following Red Cross message to his elder daughter Marianne at 4 Stanley Street, Sydney, from his house at 27 Pariser Strasse in Berlin: "I will probably be changing my apartment soon. Your letter from February has arrived. Greetings to my siblings."[141(t)] The first sentence of this carefully worded message might well have been a coded implication that Byk was expecting soon to be transported "to the East." This was the last message that Marianne received from her father. As a child, he had been very close to his sister Alice and brother Walter who had emigrated to South Africa and Palestine, respectively. It was clear he wanted them also to receive his last message.

At the beginning of June 1942, at the age of 64, Byk was taken from his home to a deportation point in Berlin. The influential Adolf Windaus, who was awarded the 1928 Nobel Prize for Chemistry for his synthesis of vitamin D, tried to stop Byk's deportation but without success.[145] Byk was transported first to Theresienstadt and then, with 756 others via Majdanek, to Sobibór on the border of Poland and Ukraine. His murder was recorded as 13 June 1942.[67] His last message to his daughter Marianne had only arrived in Geneva on 25 June 1942, and thus her father had already died when she received it. Sobibór had only been established as an extermination camp one month previously. Between 170,000 and 250,000 people were murdered there during the war. Only Auschwitz, Treblinka and Belzec had higher numbers. Alfred Byk's possessions in Berlin of a Bechstein piano, fine furnishings, stamp collection, oil paintings and library including rare scientific books were subsequently sold at public auction with the proceeds taken by the State.[145]

In 1946, Ilse Ursell of the SPSL wrote to some of Byk's original referees asking if they had any information on his whereabouts.[140] Paul Epstein (California Institute of Technology), Richard von Mises (Harvard) and Richard Tolman (now at the United Nations Atomic Energy Commission in New York) all replied to say they had no information. Ursell also wrote to the mathematician Felix Behrend whose departure from Germany to Cambridge in 1933 had been assisted by the SPSL.[146] He had then moved to Prague in 1935, but escaped back to England via Switzerland in 1938. Like many scholars assisted by the AAC/SPSL, Behrend was interned in May 1940 as an enemy alien after the German attack on Belgium and France. He was then deported to Australia on the ship *Dunera*. This was an infamous deportation ship with very poor conditions and administration. The boat was also hit by an enemy torpedo which failed to detonate. On arrival in Australia, Behrend was interned for a further period, and the mathematicians Hardy and Whitehead both campaigned for his release. He was freed in 1942 and was appointed to a position at the University of Melbourne where he had a successful mathematical career for a further 20 years.[147]

In her letter of 10 March 1948 to Behrend about Alfred Byk, Ursell wrote, "His daughters are known to have emigrated to Australia in

November 1938. They were hoping to find posts and to apply for an entrance permit for their father at the earliest possible date."[140] Behrend managed to get in contact with Byk's daughters who were now living in Frensham, New South Wales. On 3 May 1948, he wrote to the SPSL: "They advise me that Prof. Byk was deported from Berlin on June 13 1942 to an unknown destination and did not return."[140]

Marianne and Hilde Byk settled in New South Wales where Hilde married Frederick Osborne. In the 1950s, they were compensated by the German authorities for the confiscation of their father's properties and library. They both moved to the UK in the 1980s to be closer to Hilde's daughter Rosemary.[136]

On 1 February 2023, a commemorative plaque in honour of Alfred Byk was dedicated at Olivaer Platz, Berlin, opposite house number 6. This is very close to the original site of his home at 27 Pariser Strasse. The German Physical Society organised the plaque with a donation from Byk-Chemie AG, the modern name of the company set up by Alfred Byk's father Heinrich.[148]

2.5 Herbert Pese

Herbert Pese (1899–1943) was another physical scientist whose case was considered by the SPSL. He was born on 7 September 1899 to a Jewish family in Gleiwitz, Upper Silesia, which was then in Prussia but is now in Poland. His family moved to Breslau which was close to the border with Poland. There, he undertook research on acoustic vibrations and membranes with Professor Erich Waetzmann from 1924–1926. He then worked in the Physics Institute in Breslau with Professor Clemens Schaefer on optics, power fields and colour theory.[149] He completed his dissertation in 1930 on "Contributions to the Foundation and Application of Colour Theory." In the same year, he also published with Schaefer on colorimetry in the *Physikalische Zeitschrift*.[150]

Following the new laws of 1933 imposed by the Nazis, Pese was dismissed from the University in Breslau and was unable to continue research. His supervisor Schaefer contacted Max von Laue and stressed the urgent need to find funds to enable Pese to continue his promising research on colour mixing.[149] To make ends meet, Pese had to obtain employment as a forest worker and also taught mathematics and science in a Jewish school in Breslau. Schaefer wrote on 9 September 1935 to the SPSL that Pese had invented a new colour mixing apparatus on which Schrödinger reported:

> The apparatus is most thoughtfully invented and appears to be a decided improvement and simplification on the Helmholtz apparatus. The selection of fixed wavelengths and intensities is a decided superiority. Also, the omission of a Nicols prism is to be welcomed, as is the use of mixed indecomposed white for saturation compensation.[151]

Some 15 years previously, Schrödinger had briefly been a professor in Breslau where he had done original research on colour theory a few years before introducing his great work on wave mechanics.[4] Colour theory, however, unlike quantum mechanics, was not an area of major interest in UK laboratories and no professors came forward to offer Pese a position.

Breslau, which is now Wrocław in Poland, was then a city in Germany which had a large Jewish community. Unemployment was high in the 1930s and Breslau had one of the largest contingents of members of the National Socialist party.[152,153] On the Kristallnacht of 9 November 1938, the synagogue in Breslau, one of the biggest in Germany, was destroyed by fire and many Jewish people were attacked.[152] Just one day earlier, a letter had been sent to the SPSL from Dr. Lux of the Breslau synagogue community, commenting very favourably on Pese's teaching in electrical engineering.[151] Pese was arrested and held in the Sachsenhausen concentration camp from 15 December 1938 to 20 January 1939.[154] On 8 December 1938, the SPSL received a letter from Paula Morawski in Breslau with several supporting documents about Pese and saying in guarded language, "He is not able to write to you at the moment."[151(t)] Following his release from Sachsenhausen, Mrs. Steffi Grabowski, another associate of Pese who lived in London, then wrote to the SPSL on 20 March 1939, saying that "If Dr. Pese is not found a position to enable him to emigrate to England or the Dominions then there is a strong possibility he will be returned to the internment camp."[151(t)]

In due course, with more than 1400 others from Breslau, Pese was transported on 4 March 1943 to Auschwitz where he perished or was murdered.[149,154] On 31 January 1948, Ilse Ursell wrote to Mrs. Grabowski for information on Pese's fate, saying that she "very much hoped Dr. Pese managed to get away from the continent on time." Mrs. Grabowski (now with the surname of Granby) responded that she "did not have any news."[151]

2.6 Erich Lehmann

Erich Lehmann (1878–1942) came from Berlin and was the official photochemist of the Germany army in World War I.

Lehmann was born on 9 August 1878 in Berlin to a Jewish family. He studied chemistry at the University of Berlin and then worked as an assistant to Adolf Miethe at the Photochemical Institute of the Technical University of Berlin and became an associate professor.[155] Miethe had also been an associate of Alfred Byk through a mutual interest in photography.[136] Lehmann worked on numerous innovative projects. He collaborated with Erich Ladenburg in experiments on the spectrum of ozone, and in 1906 they published on this topic in the prestigious *Annalen der Physik*.[156] With Miethe, Lehmann developed methods to make artificial gemstones,

Fig. 2.14   Erich Lehmann.

which led to some private income from the German Gemstones Society.[155] He travelled with Miethe to Egypt where they observed the UV spectrum of the sun through the clearest atmospheric conditions.[157] In 1908, he also published in the *Zeitschrift für Physikalische Chemie* on the properties of photographic compounds.[158] His thesis on this topic qualified him to teach as a professor. He also lectured at the military technical academy. His expertise on the chemistry of photography was put to use in World War I where he was the official photochemist of the army and the technical adviser to the War Department on photography.

After the war, he was in much demand as a consultant on the rapidly evolving fields of colour photography and cinema technology, and he published several patents related to these areas. In 1922, he became a non-official Associate Professor of the Technical University of Berlin (see Figure 2.14).[159] He published a paper in the *Berichte der Deutschen Chemischen Gesellschaft* in 1925 on the isomers of stilbazoles which are compounds used in photography.[160] In 1927, his professor Miethe died and Lehmann was appointed as his successor two years later. He also published papers on the physical chemistry and synthesis of organic molecules of photochemical interest. In 1930, he was made the first director of the German Research Institute for the Graphics Industry and also became a consultant on cinema fire protection.[155]

After the Nazis came to power in 1933, Lehmann's professorship was terminated even though he had made important contributions to the armed forces in World War I. He was also forced to resign from his Board Membership of the German Society for Photographic Research even after an appeal to the authorities was made on his unique experience in the science behind cinema technology.[155] After 1933, he was able to continue some of his consulting work that provided a limited income, and he also had some support from family members. Consequently, he did not consider emigration initially.

On 10 March 1939, an application to the SPSL was received in Lehmann's name.[161] This was sent by his cousin Dr. Victor Lehmann who lived in London and who stated that Erich Lehmann was "one of the leading German authorities in photochemistry and cinematography, and that from 1925 to 1933 he was President of the Deutsche Kinotechnische

Gesellschaft. During the war he was the technical advisor to the war ministry on all matters concerning photography and was photochemist to the German Air Force, a unique post which was created for him." Victor Lehmann had been a judge in Berlin and was well respected through his work on the Arbitration Tribunal for the Weimar Republic.[162] The referees listed for Erich Lehmann were impressive and influential. They included Professor Charles Fabry (the optical physicist from Paris), Rudolf Ladenburg (who was now in Princeton), Dr. S. E. Sheppard (Director of Kodak Research in Rochester, USA) and Professor Thomas Slater-Price (President of the Royal Photographic Society).

The application of this talented man to the SPSL in March 1939 was a very late one and, despite the eminence of his referees, his photography-related experience did not fit in with the more conventional university-based contacts of the SPSL and there was no time to establish industrial employment. Victor Lehmann was informed by Tess Simpson that "the conditions are now extremely difficult. We would advise Professor Lehmann to get in touch with his contacts in the USA, as prospects in that country are better than in Europe."[161] One of Lehmann's sisters lived in Montreux, Switzerland, and she contacted a friend in Chicago, Numa Lachman, to see if she could help find a position for her brother in the USA. Lachman contacted the Emergency Committee in New York on 10 February 1941 about Erich Lehmann's case. However, the response was, "We are not able to take any action in these cases until an American college or university has been interested in extending a call to the scholar to become a member of their faculty."[163]

With the start of the war just six months after his unsuccessful application to the SPSL, things got very difficult for Lehmann and his relations in Germany. His other sister Margarethe Jacoby took her own life in Berlin on 19 November 1941.[164] Lehmann had stayed in Berlin, and with the mass deportations of Jewish people from that city in the second half of 1941, he also took his own life by hanging on 11 January 1942 in his apartment of 6 Carmerstrasse, Charlottenburg.[165] He was found by his housekeeper. This was just two months after both his sister and Marie Wreschner had also taken their own lives. Shortly after his suicide, the entire inventory of his apartment, which included several valuable items such as oriental carpets and Persian, Egyptian and Indian pottery, was auctioned for the State in the same way as was carried out for the possessions of Alfred Byk.[155]

On 24 March 1947, Ilse Ursell wrote to Fritz Weigert who had been named as a referee by Lehmann and had himself been assisted by the SPSL to come to the UK from Leipzig in 1934. Weigert had applied photochemistry methods in cancer research and was appointed as director of the Physiochemical Department of the Cancer Research Institute at Mount Vernon Hospital, Northwood, near London. Weigert replied that he had no information on Lehmann and suggested contacting Arnold Weissberger at Eastman Kodak in the USA. Weissberger replied on 1 May 1947: "I am afraid the news about Professor Lehmann is very bad. He committed suicide in 1940 or 1941. This information comes from Dr. Staude, a former pupil of Professor Weigert. Staude is now teaching at the University of Leipzig and at the Technische Hochschule in Dresden."[161] This was the last communication the SPSL received on Erich Lehmann.

CHAPTER THREE

Physics and Chemistry Survivors

3.1 Paul Dreyfuss

There were some scientists who Ilse Ursell at the SPSL made enquiries about after the war and who did manage to survive. On 6 February 1948, Peter Pringsheim, who played such a central role in the careers of several scientists discussed in this book, wrote to inform Ilse Ursell that he had heard that Dr. Duschinsky had been killed and "the same has probably happened to Dr. Dreyfuss whom I met in a French concentration camp, but have not heard anything since."[103]

Paul Dreyfuss (1906–1977) was born on 4 December 1906 in Elberfeld, Germany, to a Jewish family. His father Moritz came from the

Fig. 3.1   Paul and Irène Dreyfuss in their Belgian Aliens Registration Certificates, 1937.

Palatinate region of Germany and his mother Laura (née Braun) was born in Essingen. Dreyfuss was an organic chemist and wrote his doctoral dissertation at the University of Bonn under the supervision of Paul Pfeiffer. Pfeiffer had taken the chair of August Kekulé who famously claimed to have discovered the symmetrical planar structure of benzene in a dream. Dreyfuss had been funded by the chemical company I.G. Farben and he did research on new ethylene-based dyes.[103] Research into new dyes was going to be the main feature of his career, which turned out to be quite adventurous.

Early in 1933, Dreyfuss was appointed to a research assistant post in Bonn but was quickly fired from that position under the new laws in Nazi Germany in May 1933. From his family home at Königstrasse 114, in Wuppertal-Elberfeld, he wrote to the AAC on 1 June 1933.[103] The AAC had already received a letter from Professor Robert Haworth stating, "He would be pleased to have Dreyfuss in his laboratory (in Durham) but could not provide financial assistance." Dreyfuss also wrote to Lord Melchett, a Director of ICI: "Owing to the political conditions I have lost the situation of Liebig assistant which had been granted me by Prof. Duisberg (I.G. Farben) for excellent work. There is no possibility and no prospect for a Jewish scientist in Germany what so ever now."[103] On 7 February 1934, Dreyfuss received a reply from the AAC suggesting that he may be suitable for an appointment at the British Ochre and Oxide Company "on fine colour manufacture with special reference to inorganic pigments." However, by this time, Dreyfuss had moved to new positions, first at the University of Cagliari in Sardinia and then at the University of Catania in Italy.[166] There he married Irène Reinmann, a German national from Freiburg, on 2 May 1936.

Dreyfuss had also applied to the Emergency Committee in Aid of Displaced Foreign Scholars in New York in 1934.[167] With the expansion of the chemical and dye industry in the USA, there was some interest in his case including from the Rockefeller Institute in New York. In 1938, a post became available at the University of Kansas City and his papers were sent there. The same happened to an industrial position in Connecticut associated with Eduard Färber; Färber became well known both in the chemical industry and in writing on science history in the USA. Also in

1938, some positions became available in the Philippines and Dreyfuss's papers, with him labelled a specialist in "dye materials," were also sent there to the President of the Jewish Refugee Committee.[167] It seems that other scientists who were already present in the USA at that time were given priority. However, as we discuss below, an attractive opening was soon to become available for Dreyfuss in the USA.

With anti-Semitism growing in Italy, Dreyfuss and his new wife moved first to the University of Padua and then to Brussels, arriving on 9 November 1937 (see Figure 3.1).[168] He worked as a research chemist in a laboratory established by Bela Gaspar, a Hungarian chemist, who had invented and patented a novel colour film process in Berlin before escaping from Nazi Germany in 1933.[169] His invention used a three-colour process on a single strip of film. The colour image was formed from dyes already deposited on sensitised paper, which were then destroyed selectively by light. This "Gasparcolor" technique was used in cult colour animation movies by the iconic filmmakers Oskar Fischinger, Len Lye and George Pal.[170] Agfa, the photographic film company which was part of the chemical company I.G. Farben and was close to the Nazi regime, stole Gaspar's technology and then subsequently objected to his patents for many years. This forced Gaspar to move his company to Belgium and also a subsidiary to London. He managed to maintain an office in Berlin until 1935 when his employee Mr. Burmester, who was also Jewish, and his wife tragically took their own lives by jumping under a train.[170]

Paul Dreyfuss's parents, Moritz and Laura, also came to Belgium. Amongst several reports, the Belgium police files record that Dreyfuss was arrested by the Brussels police for possession of an illegal defensive weapon and fined 50 Francs on 11 October 1938. As an immigrant requiring permission to stay in Belgium, he ran the risk of being deported.[168]

Being a male German national, like Pringsheim and Wasser, Dreyfuss was arrested by the Belgium authorities in Brussels on 10 May 1940 when the German army invaded, and was sent to the St. Cyprien internment camp in the South of France where he met Pringsheim.[91,103] Figure 3.2 shows Dreyfuss in the camp and also the very cramped living

Fig. 3.2   Paul Dreyfuss, second row on the left, in the St. Cyprien internment camp in the hot summer of 1940 (upper photograph). The lower photograph shows the crowded accommodation in the camp.

quarters in St. Cyprien.[171] After a short period, Dreyfuss was transferred to the camp in Gurs, near Perpignan and then Camp-des-Milles near Aix-en-Provence. However, by then, Bela Gaspar had moved to Hollywood in Los Angeles, Southern California, where his Gasparcolor method for producing colour photographs and films was of interest to the burgeoning movie industry.[169] He wanted to bring his team of inventive chemists from Belgium to California. Accordingly, Gaspar arranged for a position for Dreyfuss in his company, which provided the essential employment he needed for a visa to move from France to the USA. His wife Irène Dreyfuss had not been arrested in Belgium and she managed to make the difficult journey in 1941 from Brussels across war-torn France to Marseilles. She recalled being stopped at the Belgian border by a sympathetic German guard who was from her home town of Freiburg. He should have barred her progress but instead prepared a statement saying she had come from Marseilles and he was just returning her there.[166,171]

After spending many hours at the American Consulate in Marseilles, Irène managed to obtain visas for her husband and herself to go to the USA.[166] They illegally crossed the border to Spain and after travelling to Portugal, took a boat, the *Nyassa*, from Lisbon on 14 July 1941 arriving at Ellis Island, New York, on 15 August after stopping in Casablanca.[58] They then travelled across the USA to arrive in Los Angeles on Labour Day, 1 September 1941.[171] What a contrast California must have been to war-torn Europe. Paul and Irène Dreyfuss's daughter, Judy, had been conceived in France and was born in Los Angeles on 12 February 1942, and her brother Michael was born four years later.[171]

Most unfortunately, Paul Dreyfuss had to leave his parents behind in Belgium. His father Moritz died of diabetes there on 25 December 1941. His mother Laura was transported on 9 October 1942 from the Mechelen internment camp near Breendonk on Transport 13 to Auschwitz, where she did not survive.[67,172]

Bela Gaspar also provided a job in California for another of his refugee employees, Walter Michaelis, who was born in 1902 and was from Bleicherode in Germany. He had been taken from Belgium to St. Cyprien at the same time as Dreyfuss (see Figure 3.3).[91] Like her friend Irène Dreyfuss, Hella, the wife of Michaelis, was left in Belgium and

Fig. 3.3   Walter Michaelis in his Belgian Aliens Registration Certificate.

made the difficult train journey with their son George to be close to her husband in the South of France. Around this time, Bela Gaspar even arranged for a patent on a process for producing photographic multicolour pictures to be assigned to his new American company Chromogen with the inventor Walter Michaelis having the address "Camp de St. Cyprien".[173] Walter Michaelis with his wife and son escaped from France to Lisbon and then to New York in May 1941.[58] They were joined by his younger brother Arthur Michaelis who had also been interned in St. Cyprien. Sadly, Walter Michaelis died in Los Angeles just three months later, but his wife Hella continued to publish patents of his work posthumously through Chromogen.

Despite the fact that Gaspar was responsible, essentially, for saving the lives of several refugees and their families, he was not popular as an employer as he paid very low salaries. It was stated that "Gaspar treated Paul (Dreyfuss) very badly."[166] Another of Gaspar's employees in Brussels, Paul Goldfinger, as discussed later in this book, was never paid at all. Times were hard for the young Dreyfuss family in California and Irène had to get a job sewing aprons to make ends meet.[166] The photographer Richard C. Miller used the Chromogen technology to take iconic photographs.[174] He became celebrated for his vivid colour photographs of famous film stars including Norma Jean Dougherty (later called Marilyn Monroe) in the 1940s and 1950s. Miller even photographed Gaspar and Dreyfuss in their laboratory (see Figure 3.4).

Fig. 3.4   Dr. Bela Gaspar (top) and Dr. Paul Dreyfuss (bottom) in the Chromogen Laboratory, Hollywood, ~1944. Photographs by Richard C. Miller.

The research work of Dreyfuss in California with Gaspar was initially successful and in 1948 they published a joint patent on new acid azo dyes that gave brilliant blue colours.[175] However, Gaspar was not able to compete with the rival Technicolor technology and his company went bankrupt in 1951.[169] Dreyfuss then had a series of appointments with the chemicals company Ciba. He moved to Basel in Switzerland in 1951 but was back in the USA in 1954 with Toms River Chemicals in Cincinnati. In 1957 he returned to Basel. He continued to publish patents on dyes for use in colour photography. He also played a key role in transferring the Chromogen technology back to Europe, where it had been initiated by Gaspar in the first place in Germany, and the company Cibachrome was established.[169,176] By this time, the Chromogen patents had run their course. The Cibachrome colour prints had a polyester base with 13 layers of azo dyes and this prevented them from fading and discolouring. This has made such photographs collectors' pieces. Cibachrome was also subsequently called Ilfochrome after a merger between Ciba Geigy and the British Ilford photography company. The technology was still being used until 2012, by which time digital innovations had revolutionised colour film.

Paul Dreyfuss died in Basel in 1977 at the age of 71, just four years after Bela Gaspar. He led a colourful life in more ways than one. The SPSL was unaware of Dreyfuss's move to California and his subsequent career. Its final record on him stated, "He was last heard of in a French internment camp in 1940."[103]

3.2 Three Chemical Physicists

Vladimir Lasareff, Paul Goldfinger and Boris Rosen can be considered as a close group of scientific refugees as they were all non-German nationals from Jewish families (Lasareff and Rosen were born in Russia and Goldfinger in Romania). All three came to work with leading groups in Berlin where they researched problems in chemical physics.[107] Then, following the events of 1933, they were all employed at the University of Liège in Belgium in the laboratory of Victor Henri. Henri had done pioneering work on the chemical kinetics of enzyme reactions and had then shifted his research interests to spectroscopy which was providing the fundamental parameters on molecular structures and energetics needed to understand chemical processes. Henri had Russian parents but was born in France. He had positions in Moscow and Zurich, where the duration of his stay overlapped with that of Schrödinger at the University of Zurich in the 1920s. During this period, Henri was the first to discover the effect, dubbed predissociation, of the broadening of spectral lines of molecules in temporarily bound states.[177] After a brief transfer to Marseilles, Henri then took up a chair at the University of Liège. He realised the opportunity of having the three exiles and experts from Berlin in the expanding field of molecular spectroscopy in his laboratory with two of them coming from Russia, the country of his parents.

Henri's grant for these scientific refugees was to expire in 1935 and, at that time, foreigners could not be appointed to permanent academic positions in Belgium. There was also some concern from the Belgian authorities that the influx of refugee scientists was taking away employment from their own nationals. Therefore, Lasareff, Goldfinger and Rosen all applied in 1935, with Henri's strong support, to the AAC for assistance. However, with three applications from refugee scientists with publication records all in the same field, the challenge to the AAC to find suitable positions was too great. All three were working on detailed experimental spectroscopy of molecules. This was a relatively new field which was still to take off in a major way in the UK as we have seen in the discussion on Fritz Duschinsky.

Despite the financial difficulties that had been expressed in letters to the AAC, Lasareff, Goldfinger and Rosen continued to get some support from the University of Liège after 1935 and were to stay in Belgium during the war. An emotional plea had been made to the Belgian immigration authorities on 23 July 1935 from the Rector of the University to allow the three refugees to stay in Belgium:

> I have the honour to inform you that indeed Messrs. Goldfinger, Lasareff, and Rosen are the first two attached respectively to the physical chemistry laboratory, the third to the astrophysics laboratory of the University of Liège. Their stay in Belgium was authorized at my request.

The remuneration they receive, two thousand francs per month, is very modest, since they are all married and fathers of a child. To collect the necessary funds, I appealed to the Francqui Foundation, to the British committee for the aid of Jewish scholars and to my colleagues at the University of Liège, who, in a fine spirit of solidarity, brought me their contribution. In favour of prolonging the stay of these three refugees in Belgium, there are two reasons. First, there is a humanitarian reason. Victims of an abominable persecution Messrs. Goldfinger, Lasareff and Rosen are devoid of all resources. They don't even have a homeland. What would become of them if they were denied asylum as our country kindly granted them for two years?

There is also a reason of utility for the University of Liège. Our three refugees are valuable scholars and their stay in our laboratories has been extremely fruitful. Students of Einstein and Haber, they put their theoretical and technical knowledge at the disposal of our students, for the greater benefit of all of them. I would add that all have shown themselves to be irreproachable both from the point of view of conduct and character. One of them Mr. Lasareff has had the misfortune to lose one of his children since his arrival in Liège and I admired the courage with which in this terrible ordeal, despite the insecurity of his situation, Mr. Lasareff continued his work.[178(t)]

Significantly, none of the three were German nationals and thus, unlike Pringsheim, Wasser, Dreyfuss and Michaelis, they were not interned by the Belgian authorities when the German army invaded Belgium in May 1940. After several dangerous incidents which are described in the following sections, Lasareff, Goldfinger and Rosen all survived the war and went on to have impressive careers in Belgium afterwards.

3.2.1 Vladimir Lasareff

Vladimir (Wladimir) Lasareff was born to a Jewish family on 25 March 1904 in Russia in what was then called St. Petersburg. His family moved to Finland following the Russian Revolution in 1918 and then on to Berlin where Lasareff studied at the University.[107] He completed his dissertation in 1927 on the broadening of vibrational-rotational spectral lines in gases. From 1930–1933 he worked in the laboratory of Professor

Fig. 3.5   Vladimir Lasareff in his Belgian Aliens Registration Certificate, 1933.

Freundlich at the Kaiser-Wilhelm-Institut für Physikalische Chemie und Elektrochemie in Berlin-Dahlem where the duration of his stay partly overlapped with that of Marie Wreschner.[107] During this period, he also became friendly with Albert Einstein who showed an interest in his career over several subsequent years. Einstein had left Germany for the USA in 1932 but returned briefly to Europe in 1933 after Hitler came to power to warn of the danger. Einstein then corresponded with Lionel Ettlinger, a wealthy businessman based in New York, who was attempting to set up an international group based in Geneva to oppose Hitler. On 4 May 1933, Einstein wrote to Ettlinger asking if he could assist Lasareff in obtaining a visa to move to Belgium.[179] Shortly after this, Lasareff, together with Goldfinger and Rosen who had also lost their positions in Berlin under the new Nazi laws, moved to the temporary appointments at the University of Liège in the laboratory of Victor Henri (see Figure 3.5). Lasareff's application to the AAC was received on 8 March 1935 in which he described himself as stateless.[180] Einstein then wrote to the AAC from Princeton on 16 February 1935:

> Two years ago, the University of Liège took on three capable young physicists, one of whom I know very well from Berlin, Dr. Lasareff, who

works with Professor Henri. The University of Liège has hitherto raised the very modest salaries privately, but the available funds will only last until June 1, 1935. The Rector of the university has written to me the enclosed letter, in which the matter is clearly explained. Unemployment among young physicists is so bad here in this country that one dare not seek employment for foreigners unless they have a world-wide reputation. The University of Liège would like to keep them and living there is particularly cheap as Belgium is one of the countries with the lowest prices. What could be done in this matter? Can you perhaps tell me a place where I could intercede for these three?[180(t)]

The other two physicists referred to here by Einstein are Goldfinger and Rosen who had published in Belgium with Lasareff on the dissociation energies of diatomic molecules.[181] Following the inability of the AAC to assist the three chemical physicists to find appointments in the UK, Lasareff continued with his research in Liège and moved into the biochemistry area, publishing works on the adsorption of molecules on surfaces and protein interactions.[182] His research continued for a short period after the invasion of Belgium by the Nazis in May 1940.

However, things changed when Germany declared war on Russia on 21 June 1941 and the Nazis in Belgium started to round up Soviet citizens.[183] This included Isaac Lasareff, a doctor and Vladimir Lasareff's father, with whom he had moved to Belgium from Berlin in 1933. Isaac Lasareff, aged 64, was interned in the Belgian prison at Breendonk near Mechelen about half way between Antwerp and Brussels.[172,183] Originally a military fort, Breendonk had been requisitioned by the SS as a prison camp for political dissidents, Jews and Resistance members.[184] It was to become a place with a notorious reputation.[172,184]

On hearing about his father, and realising he would not survive long at Breendonk, Vladimir Lasareff at once went to the Brussels headquarters of the SS to complain. Then, quite extraordinarily, the SS informed him that he would take his father's place. He was given the same prisoner number, bed and uniform as his father who was released in an unrecognisable state after just eighteen days in the camp.[172]

At Breendonk, Vladimir Lasareff undertook forced labour to which he was not suited and which nearly resulted in a foot amputation. He also witnessed the brutal death of a friend Israel Neumann who was the subject of a savage attack by the pet Alsatian dog of the camp commandant Philipp Schmitt. Eventually, in a very ill state, Vladimir Lasareff, who had been sent to a hospital in Antwerp to recuperate, was allowed to leave the camp in November 1941 during a temporary lull in the hostile atmosphere of Breendonk when several prisoners were freed.[172] He was extremely fortunate as transportations from Breendonk to camps such as Mauthausen, Buchenwald and Auschwitz started a few months later in May 1942 as the war accelerated and the Belgian Resistance stepped up their activities. One of those transported to Auschwitz from Mechelen, the deportation centre for Breendonk, was Laura, Paul Dreyfuss's mother.

In the meantime, a concerned Tess Simpson had written from the SPSL to Einstein on 29 October 1942:

> We are anxious to obtain information if possible about the refugee scholars registered with this society who, when we last heard of them, were established in countries since occupied by the enemy. We are hoping that many have succeeded in escaping to a neutral country or to America. One of these scholars is Vladimir Lasareff, physicist, who last wrote to us from Liège. He gave your name as a reference and we were wondering if by any chance you have more or less recent news of him.[180]

Einstein replied on 16 November 1942: "To my sincere regret I have no news of Lasareff since the invasion of Belgium. I have a very good opinion of Dr. Lasareff and would be greatly relieved to know that he is safe."[180]

Lasareff had not published scientific works since 1939. However, in due course, the situation improved and, following the liberation of Belgium in September 1944, Einstein wrote to Ettlinger on 16 December 1944 to say, "I'm sure you'll be happy to know that our friend Doctor Vladimir Lasareff has happily overcome this terrible time."[185(t)] However, Victor Henri, the supporter of Lasareff, Goldfinger and Rosen, had died in La Rochelle in France on 21 June 1940 following the capitulation to Germany. He had moved to Paris to undertake military research with the CNRS and had a serious lung infection. This may have caused Lasareff to show an interest in moving to the USA and he had corresponded with both Ettlinger and Einstein about the possibilities near the end of the war. Einstein, however, was somewhat negative and on 15 February 1945, he wrote to Ettlinger from Princeton:

> I too have much sympathy and respect for the scientific skills of Dr. Lasareff. If he persists in wanting to emigrate to America, I would also like to help him. However, I am firmly convinced that the prospects for him here after the end of the war would be very unfavourable ... it seems to me that the prospects must be better in European countries, despite the amount of destruction. Lasareff has certainly made friends in Liège who will do their best for him.[186(t)]

On 24 May 1946, the UK Search Bureau for Germans, Austrians and Stateless Persons in Central Europe wrote to the SPSL to say that Vladimir Lasareff was still living in Liège and gave his address as 47, rue de Serbie.[180] This good news was forwarded by Ilse Ursell to Einstein who was already aware of Lasareff's situation. Ursell then wrote to Lasareff on 29 July 1946 saying, "how glad we are that you are alive and working." On 15 August Lasareff replied:

> Greatly pleased to feel your sympathy. I thank you heartily for your kind attention. Indeed, I survived to all the dangers of those trying years. After having been imprisoned in the concentration camp of Breendonk for five months, I left Liège in 1942 and then was a member of Ardenne F.I. Forces until Liberation. From October 1944 until August 1945 I was serving voluntarily with US forces in Belgium and Germany. In autumn 1945 I have resumed my scientific activities at the University of Liège.[180]

Lasareff stayed in Belgium and was made a naturalised citizen on 11 December 1948. In that year, he published works in physical chemistry from Liège.[187] He then transferred into administration in the atomic energy field and by 1958 had become the Service Chief at the Centre d'Étude de l'Énergie Nucléaire (CEN) in Mol in Belgium.[188] He moved on to become the Secretary of the European Atomic Energy Society, a post he was still holding in 1963.[189]

Over 3,500 prisoners were held in the Breendonk camp during the war, of whom over 300 were known to die in the camp itself while over 1,700 were sent to their deaths in the camps "in the East"[172]. In 1946, war crime trials were initiated against the members of the German and Flemish SS who were involved in brutality at the camp at Breendonk. Sixteen were sentenced to execution and two of these were commuted on appeal. The camp commandant Philipp Schmitt, who was involved in the brutal death of Israel Neumann which was witnessed by Vladimir Lasareff, was tried in 1949 and found guilty of the inhuman treatment of prisoners at Breendonk. He was sentenced to death and was executed by a firing squad on 8 August 1950. He was the only German to be executed in Belgium for war crimes and was the last person to receive capital punishment in Belgium.[172] Breendonk is now a National Monument that preserves the memories of the Holocaust in Belgium.

3.2.2 Paul Goldfinger

Paul Goldfinger (1905–1970) was born on 10 January 1905 in Szászrégen, Transylvania, which was a town then in Hungary and subsequently Romania. He was one of three brothers who went on to distinguished careers. His father was Oskar, an innovative businessman, and his mother was Regine, née Haimann. As a child in a well-off family, he went to school first in Budapest and then in Lausanne and he was fluent in the English, Hungarian, German and French languages. His father was born in Poland and Paul Goldfinger had a Polish passport. He studied at the ETH in Zurich where he obtained his PhD in organic chemistry in 1929 with Professor Richard Kuhn who went on to win the Nobel Prize for Chemistry in 1938 for his work on carotenoids and vitamins. Goldfinger then went to Berlin in 1929 to the Kaiser-Wilhelm-Institut für Physikalische Chemie und Elektrochemie in Berlin-Dahlem as an assistant with another famous Nobelist, Fritz Haber.[108] In that institute, he also interacted with Farkas, Harteck, Kallmann, Freundlich and Polanyi. In 1931, he married Kate (Kathe) Deppner in Berlin. She was a professional dancer who had been trained at the Max Reinhardt School for Film and Theatre. She was not Jewish. She had previously been

Fig. 3.6   Paul and Kathe Goldfinger (née Deppner) in their Belgian Aliens Registration Certificates of 28 October 1939, subsequently stamped with "Juif-Jood" in his case.

married briefly to Wolfgang Pauli who said, in his typically brutal style, "Had she taken a bullfighter I would have understood — with such a man I could not compete — but a *chemist* — such an average chemist!"[190]

Following his move to Belgium, Goldfinger worked on the electrolytic preparation reactions of heavy water and published with Lasareff on this topic.[191] He also published with both Rosen and Lasareff on the dissociation energy of carbon monoxide.[181] His daughter Marianne was born in Liège in 1934. Georges, his younger brother by six years, was also given a temporary position at the University of Liège. Georges' research area was polymer science and he immigrated to the USA before the war started. He initially worked at the Brooklyn Polytechnic with Herman Mark, who himself was a scientific refugee from Vienna, before obtaining academic positions at the University of Buffalo and then at North Carolina State University. He died in 1984.

In Paul Goldfinger's application to the AAC dated 4 February 1935, his referees included Kuhn, Haber, Freundlich, Weizmann, Franck and Henri.[192] Understandably, he did not name Wolfgang Pauli as a referee. By this time, Kuhn had moved to Heidelberg and had become a staunch supporter of the Nazis and had even denounced Jewish colleagues. He became president of the German Chemical Society and, during the war, developed the lethal chemical nerve agent soman under the direction of the German Army Ordnance Service.[193] Kuhn's reference for Goldfinger's AAC application, however, was strongly supportive stating, "I have to particularly emphasise the theoretical independence with which Mr. Goldfinger dealt with many of the physical chemistry questions in his frontier work and he has an excellent overview of the most diverse problems in chemistry."[192(t)] Also, James Franck wrote to say:

> Dr. Paul Goldfinger is well-known to me by many discussions about questions on auto-oxidation processes together with Professor Haber. I had always the impression that Dr. Goldfinger was one of the most able co-workers of Professor Haber and also his papers, which I read, confirm this impression. I was struck by his independent mind of thinking and his ability in the experimental respect.[192]

Funds were getting tight and the Belgian Office of Information reported that Goldfinger was receiving financial support from his older brother (Ernö) in London.[178] Then, in October 1936, Goldfinger wrote to Freundlich to say that he had found an industrial position in Belgium and there was an opening for a colleague with experience in organic chemistry of azo and cyanin dyes. This was communicated by Freundlich to the SPSL.[192] The name of the company was not indicated but it was that of Bela Gaspar who had also employed Paul Dreyfuss and was working on colour photography. Like Goldfinger, Gaspar was originally from Romania. This change of employment had been confirmed in a letter of 23 November 1935 from the Rector of the University of Liège to the Minister of Justice in Brussels.[178] In 1939, Goldfinger visited his father in Budapest with his daughter and worked for a short period in his father's cellulose business. However, this did not work out and they soon returned to Brussels. On the way back, he had a scare when he was arrested for a few hours by immigration officials in Nazi-controlled Vienna. With the political situation in Europe deteriorating quickly, Goldfinger then arranged for his parents to come to Brussels.[194]

The next communication from the SPSL on Goldfinger was a letter on 20 October 1942 from Tess Simpson to James Franck at John Hopkins University, Baltimore, USA, stating concern about refugee scientists in occupied countries. She mentioned Paul Goldfinger and Boris Rosen. The well-connected Franck replied that Dr. Goldfinger "still had his position in a factory in Brussels in 1941."[192]

Then, in 1947, Ilse Ursell noticed a paper by Goldfinger and Rosen in *Nature* and wrote to Goldfinger to say she was delighted he had survived the war and said that the SPSL would be pleased to have a report of his activities.[192,195] Goldfinger replied in dramatic terms, from the University of Nancy in northeast France, on 20 April 1947:

> Actually, I have had an extraordinary luck in surviving the war. Before the war I worked as a chemist in the Industrial Research Laboratory of a Romanian Jew called Bela Gaspar. He had left Belgium in October 1939 and gave full powers to a man whose first deed, after I had tried in vain to escape to Great Britain in May 1940, was to denounce me to the Germans, who imprisoned me for three weeks. I owe my liberation to the courage and energy of my wife. Gaspar on the other hand owes me still 8 months salary, that is about 40,000 Belgian francs.
>
> After this adventure I tried to earn my life by some small chemical business and on the other hand engaged in the Belgian Partisan Army — Milices Patriotiques. Here again, I had extraordinary luck. I could continue my work until 1944 when the last of my comrades got caught by the Germans. Only two of them came back after Germany capitulated. Since October last year I am Chargé de Cours at Brussels University, for photochemistry, and Chargé de Recherche at the University of Nancy. I am engaged in research work on bond energies and free radical reactions in gases and solutions.[192]

It is clear from this letter, and the previous one to Freundlich, that Goldfinger's time had overlapped with that of Paul Dreyfuss in the company of Bela Gaspar in Brussels after he had finished working in the laboratory of Henri in Liège. However, unlike Dreyfuss and Michaelis, he did not follow Gaspar to California. Goldfinger did manage to publish a patent during this period as the inventor of a light-sensitive multilayer colour photographic material (assigned to Gaspar).[196]

From 7 May 1942, the Germans made Jewish people in Belgium wear a yellow star and Goldfinger's police file indicates that he was marked in this way (see Figure 3.6). In an interview conducted in 1999, Peter Goldfinger, the nephew of Paul, stated that his uncle walked to Dieppe with his daughter to attempt to get on a boat to England, but this was unsuccessful.[197] It was also mentioned that Paul Goldfinger used his chemical expertise to make explosives for the Resistance.[197] The apartment of the Goldfinger family in Brussels was taken over by a German collaborator and Goldfinger's daughter Marianne went to live with another family for a period. With the aid of a grandmother in Leipzig, Kate Goldfinger managed to prove that she was an "Aryan" and this enabled her and Paul to find another apartment. However, after the Germans left Belgium in 1944, the collaborator disappeared and the family had the apartment back.[194]

After the war, Goldfinger became a naturalised citizen of Belgium and returned to academia at the Free University of Brussels. He published original works on the kinetics of fundamental chemical reactions in prestigious publications such as the *Journal of the American Chemical Society* and the *Journal of Chemical Physics*.[198,199] He also established a significant reputation for himself in research in the technique of mass spectrometry.[200] Goldfinger helped to establish new scientific journals including *Chemical Physics Letters* (which was first published by *North Holland* in 1967 and which the author of this book edited from 2000–2020).[107] He continued to publish right up until his death on 25 March 1970 and overlapped with several scientists still working today. Paul Goldfinger's daughter Marianne remained in Belgium and had a career as a biologist. Her partner Jacques Jauniaux was a painter and sculptor. Her son Yves Cape is a successful cinematographer.[201] Marianne Goldfinger died on 19 May 2022 at the age of 88.

Several of Paul Goldfinger's papers in physical chemistry remain highly influential, with five having over 100 citations at the time of writing this book. To rebut Wolfgang Pauli, Paul Goldfinger was no "average chemist".

Paul Goldfinger's older brother Ernö immigrated to England and lived next door in London to Ian Fleming, the writer who is famous for his creation of the fictional spy James Bond. Ernö, a very successful architect, wanted to extend his house through the acquisition of Fleming's property. Fleming initiated a lawsuit against this proposal but lost.[201] However, he got his revenge by creating the exotic villain in the hugely popular novel that was turned into the James Bond film "Goldfinger".[202]

3.2.3 Boris Rosen

Boris Rosen (1900–1974) was born on 30 August 1900 in Russia. He first studied in Crimea but after the Russian Revolution moved to Berlin where he worked first with Pringsheim and then Haber. As stated earlier, he collaborated with Goldfinger and Lasareff on the spectroscopy of diatomic molecules after moving to the University of Liège in 1933 (see Figure 3.7).

Rosen's letter to the AAC of 23 February 1935 summarised succinctly his career up to that time:

> I was born in St. Petersburg on August 30th, 1900. My parents were Russian citizens. I lived in St. Petersburg until 1918 and there I followed the courses in a high school and got a bachelor degree. After the revolution I had to leave Russia; thus I studied at the University of Berlin and I got there my doctor degree. I spent one year as assistant of Prof. Pringsheim at the Physical Institute of the University. From 1928 to 1933, I worked as a research fellow in the laboratories of Prof. Haber in Berlin-Dahlem. In summer 1933 I had to leave Germany and in October 1933 I was given the possibility to continue my work at the University of Liège. In Liège I have spent most of my time in the Astrophysical Department, equipping a laboratory for researches connected with collisions between electrons and atoms or molecules by means of mass-spectroscopy. At the same time, I go on with my previous spectroscopic investigations on molecular bands. I am married and have one child.[203]

Fig. 3.7 Boris and Lidja (née Gabis) Rosen in their Belgian Aliens Registration Certificates of 26 September 1935.

In his reference to the AAC, Professor Henri from the University of Liège stated that Rosen had started a productive collaboration there with Professor Pol Swings on applications of spectroscopy in astrophysics. This was to be the field of Rosen's research for the rest of his remaining career, which included work at Liège during the war and afterwards. James Franck also wrote to say, "Dr. B. Rosen is known to me as a co-worker of my friend Peter Pringsheim and later of H. Kallmann. I had always a very good impression of his personality as well as his scientific ability."[203]

Although Henri had indicated to the AAC that funds in Belgium were very short, Rosen managed to continue his research at the astrophysical institute, and the Rector of the University of Liège indicated to the Belgian Authorities that private funds were used for his support.[204]

On 20 October 1942, a concerned Tess Simpson wrote to the well-informed Franck asking about Goldfinger and Rosen, with the stated hope that they had managed to escape to a neutral country or to America. Franck replied from the University of Chicago on 17 November 1942:

> The only thing I know is that Dr. Goldfinger had still his position in a factory in Brussels in 1941 and that Dr. Rosen was reinstated at his position at the University of Liège, astrophysical laboratory, beginning 1941. I fear that in the meantime their position became a very bad one. I would be very interested if you are able to find something about the fate of these men.[203]

A similar enquiry was sent by Tess Simpson to Pringsheim on 14 June 1944 saying, "About Dr. Rosen we have heard nothing at all." Pringsheim replied on 17 July 1944:

> Dr. Rosen was as late as April 1944 with his wife and children in Liège, not only unharmed but going on with his academic work and paid by the university. Dr. Swings, who was Rosen's superior at the astrophysical institute in Liège, and is since the beginning of the war in the USA, told me he is in regular correspondence with his parents still living in Liège (some friends in Stockholm transmit the letters) and they mention quite regularly that Rosen is there - I hope that this is true even now.[203]

It seems that, at some time during the war, Rosen took up a false name and had a secret existence with his wife and son in Liège.[107] After the war, Ilse Ursell wrote to Rosen on 18 September 1946 to say that she had been delighted to find an article by him recently published in *Nature* and had obtained his address in Liège from the International Red Cross.[195,203] She also communicated this good news to Pringsheim. Rosen's collaborator Professor Swings, who had made the first detection of a molecule (CH) in interstellar space, had been in the USA in 1939 when the war broke out and stayed there for its duration. Afterwards, Swings returned to his institute directorship in Liège and continued to encourage Rosen in his work on spectroscopy relevant to astrophysics. Rosen published a compilation of spectroscopic constants in 1951 which has been widely used.[205] In 1964, in a very ambitious project which was not subsequently continued, he attempted to produce an artificial cloud of ammonia with a rocket at an altitude of 242 km.[206] As late as 1970, he edited a volume on the optical spectroscopy of solids.[207] He was involved with several committees of the International Astronomy Union and was awarded the Commandeur de l'Ordre de Léopold II. He died on 2 January 1974 in his adopted hometown of Liège.

3.3 Alfred Lustig

Alfred Lustig (1908–1985) was unusual in that he was sent from Vienna to one of the first concentration camps in Poland but managed to escape to the Soviet Union, fought in the war against Germany and lived for the rest of his life in Russia (see Figure 3.8).

Lustig was born to a Jewish family in Vienna on 10 October 1908. His father was a cabinet maker. He studied physics and mathematics at the University of Vienna from 1927–1931 and his teachers included Ehrenhaft and Hans Thirring. His research dissertation was on Brownian motion. He then became a research assistant in Vienna to Ehrenhaft working on sub-microscopic matter and authored a paper with Max Reiss who also published with Emanuel Wasser.[208]

Fig. 3.8 Alfred Lustig.

Following the Anschluss in Austria, Lustig wrote almost at once, on 2 May 1938, to the SPSL requesting assistance.[209] In addition to Ehrenhaft and Thirring, his referees included Friedrich Kottler and Philipp Frank. However, as was the case with Emanuel Wasser, his research in Vienna was not of major interest to the leading professors in the UK, and most members of the Ehrenhaft school of research did not have success with their applications. Following a further pleading letter from Lustig, Tess Simpson wrote to him on 7 June 1938 in her usual polite but frank style: "We shall do all we can to help you to find a position, but it is only fair to tell you that the difficulties now are very great and that there are very few vacancies. If you have any contacts in the USA we would advise you not to delay in getting in touch with them."[209] Tess Simpson also received a pleading letter on 24 June 1938 on behalf of Lustig, and a medical researcher Benno Grossmann, from Eric Mann of Dudley, Worcestershire, who stated, "The government (in Austria) makes it as hard as possible for them to get in touch with friends or movements in other countries." Simpson responded quite firmly: "We are trying to do all we can to help them to find a suitable position, but this becomes more and more difficult. We know that conditions in Vienna are appalling and if only we could we would try and get all the scholars and scientists out. However, to invite someone to come to this country means taking indefinite responsibility for him, and this we cannot do for the hundreds who are in need."[209]

Lustig also made a formal application to the central authority in Vienna on 13 May 1938 for permission to leave Austria and stated his preference would be to move to the USA or Australia.[210] Then, in June 1938, like many other Jewish academics in Austria, he was fired from his position at the University of Vienna. The SPSL was contacted by Lustig's sister-in-law on 29 June 1939 saying he had registered with the US consulate and an American Committee was investigating his guarantees.[209] She asked if the SPSL could forward the details it had on her brother to the US German Jewish Aid Committee in London. However, this was only two months before the start of the war and it was too late.

In 1939, Adolf Hitler devised a plan to send Jewish people to the corners of the territories that had been conquered by the Nazis. An area close to Lublin and Nisko in Poland was chosen as a "reservation". This was close to the border with the region of Poland recently acquired by the Soviet Union. Adolf Eichmann was involved in making the arrangements for the transportations from Bohemia, Moravia and Vienna. The first transportation from Vienna was made on 20 October 1939.[211] Alfred Lustig was in this group of nearly 1,000 men. The train passed through Jaroslaw, Katowice, Tarnów and Kraków and took three days. On arrival, the prisoners were marched towards wooden barracks which were still under construction. The craftsmen in the group were separated to work on the barracks, while the rest of the group were ordered to keep marching. The shots and threats were so great that several prisoners escaped across the border through the River Bug into the territory under the Soviets. One of these was Alfred Lustig.[212] The organisation of this reservation was so poor that nearly 200 of the prisoners in Nisko were eventually returned to Vienna. The chaotic experience of this experimental process was noted and a much more organised and industrial plan was then formulated by the Nazis to make transportations to the east.

In the Soviet Union, Lustig, in his own words, at first "lived for two months in the Stanislav mountains of Western Ukraine."[212] He was then sent to work in Kazan and after that to Yelabuga, Tatarstan, 1,000 km east of Moscow. He initially looked after horses in a brewery but his ability for solving mathematical problems was soon discovered by students. He was then able to teach mathematics and physics at a teacher's college. After the war against Germany started on 22 June 1941, he was drafted into the Soviet Army in January 1942. He was badly wounded in the bitter fighting of November 1942. He was then called again to the infantry in April 1943 to the second Baltic front and was seriously wounded again.[212]

After the war, which he was most fortunate to survive, Lustig was awarded the "Order of Glory, III Degree" and, in 1947, the medal "For Valiant Labour in the Great Patriotic War" by the Supreme Soviet of the USSR. He returned again to Yelabuga and taught himself Russian and the local Tatar dialect. He was first married to Zinaida Zamoreva and then to Lydia Alekseevna Belyaeva with whom he had a son, Mikhail, and a daughter, Zinaida (see Figure 3.9). Both his children eventually became university teachers and Mikhail followed in his father's footsteps as a mathematician. Alfred Lustig was appointed senior lecturer in the

Fig. 3.9   Alfred Lustig with his wife Lydia and children Mikhail and Zinaida in Yelabuga, Russia.

Department of Physics and Mathematics at the Teachers' Institute in Yelabuga. He also taught students the German language.[212]

Lustig had to make considerable effort to recover the certificate providing evidence of his degree at the University of Vienna. He finally received his diploma in 1958 and this allowed him eventually to be appointed as an Associate Professor of Mathematics in 1964 in Yelabuga. His expertise had already been rewarded with the headship of the Department of Mathematics there. He also published a paper on mathematical analysis in a Russian journal.[213] He was highly regarded by his colleagues and was awarded the Order of the Badge of Honour by the Supreme Soviet for his teaching contributions.[212]

Lustig had the opportunity to return to Vienna but he declined. There is no evidence he made any reparations claims back to Austria. He died on 21 March 1985 at the age of 76 and he was buried in Yelabuga. His death was much lamented by the friends he had made there. In the words of a colleague at Yelabuga, he was "an Austrian anti-fascist, a soldier of the Soviet Army, a physicist and mathematician, an excellent teacher, a teacher of Tatarstan teachers — we must keep the memory of such a person forever."[214(t)]

After the war, the SPSL had made its usual enquiries on refugees who had made applications but had lost contact with the Society. On 4 January 1946, the UK Search Bureau for German Austrian and Stateless Persons from Central Europe wrote to the SPSL to say that Alfred Lustig, born 08 08 07 in Duernkrant, was in Vienna after being liberated from a camp. However, the alert secretary wrote back to say the birthdate and place of birth did not link this Alfred Lustig with the one on the SPSL's books.[209] In 1947 and early 1948, Ilse Ursell also wrote to several people who were referees, colleagues or friends of Lustig asking if they had information on his whereabouts, and this included Eric Mann who had written on Lustig's behalf in 1939. There were some tentative responses that he had immigrated to the USA. Then, on 3 February 1948, a postcard was received from Professor Hans Thirring in Vienna stating, "Dr. Alfred Lustig is said to be a teacher in Russia. His address is unknown."[209] Ursell then wrote to Lustig's collaborator in Vienna, Max Reiss, who was now based in Rochester, USA. On 7 March 1948, Reiss replied, "I know that at the time of the Hitler regime he fled to Russia and was working there as a chemist. Friends of mine, Dr. Fritz Isser and his sister, were in touch with him for some time but then lost contact with him."[209]

The University of Vienna retains a Memorial Book for its members who died in World War II. There, it states that Alfred Lustig was deported "to Nisko in occupied Poland on 20 October 1939 and did not survive."[215]

3.4 Giulio Bemporad

Giulio Bemporad (1888–1945) was a distinguished Italian astronomer. He remained in Italy during the war (see Figure 3.10) and was closely involved in assisting Jewish refugees escape to the USA. He died just after the end of the war in Rome.

Bemporad was born in Florence, Italy, on 3 January 1888. He was from a Jewish family and studied mathematics and astronomy in Pisa at the University and the Scuola Normale Superiore. He wrote his thesis on differential geometry at the University of Catania and graduated in 1910. He then took up a post as assistant astronomer at the International

Fig. 3.10   Giulio Bemporad in 1943.

Latitude Station at Carloforte on the San Pietro Island, facing the southwest coast of Sardinia.[216] There, he was involved in systematic observations of latitude variations in collaboration with other centres in Japan and California. This position on an island near another island was very isolated. In 1925, Bemporad moved to Naples becoming the vice-director of the observatory there and worked on the Astrographic International Catalogue. In 1932, he moved to Turin as vice-director of the observatory. He was also subsequently classified as a university professor in theoretical astronomy. Giulio Bemporad had a cousin, Azeglio Bemporad, who was also an astronomer and was something of a rival. Azeglio initially worked in Turin but then became director of the Catania Observatory.[216]

Many Italian Jews supported Mussolini's Fascist party in its earlier stages when it was opposed to anti-Semitism and Bemporad signed an oath of allegiance to the Fascist party in 1934.[216] However, following the invasion of Ethiopia by Italy in 1935, the tolerance towards minorities in Italy reduced and the rise of Hitler was admired by many of the Fascist leaders there.[217] In 1938, a Racial Manifesto was issued which stated that Italians, like the Germans, had descended from the Aryan race.

All education employees of the State were asked to declare if they belonged to the Jewish race. A nationwide census was also carried out in which the question of Jewish parentage was asked. The political atmosphere was changing quickly after the Munich Agreement of September 1938 signed by Germany, the UK, France and Italy, with Mussolini playing a major role. A law was passed in Italy on 17 November 1938 which excluded people with Jewish ancestry from public offices and higher education.[217] In the next few years, the discrimination became more severe and Jews had their properties confiscated, were interned and sent to concentration camps. As a consequence, several distinguished scientists left Italy, with Enrico Fermi being the most prominent example.[218] This exodus included astronomers.[219]

The number of Italians who applied to the SPSL was much smaller than academics from Germany or Austria. The tally of refugee scholars from different countries that were helped by the AAC/SPSL and responded to enquiries after the war was Germany (364), Austria (95), Czechoslovakia (26), Italy (19) and Poland (17).[2] Nevertheless, the SPSL was able to provide some assistance to a small number of leading Italian scholars.[220] This included the mathematician Gino Fano from Turin and his physicist son Ugo Fano who had worked in Rome with Fermi.[221,222] In due course, father and son spent most of the war in Switzerland and the USA, respectively. Another distinguished mathematician was Beniamino Segre from Bologna who obtained a position in Manchester.[223] He was one of the founders of finite geometry. However, as we have seen in applications to the SPSL of refugees from Austria and Czechoslovakia, it was much more difficult to help scholars in 1939 than in the previous years because most suitable temporary positions in the UK had already been taken.

Just three weeks after the establishment of the new laws in Italy, Giulio Bemporad was dismissed on 8 December 1938 from his position in Turin. He then wrote to the SPSL on 8 January 1939.[224] He listed 45 publications in his application on astronomical observations and their interpretations through calculations. He had an international set of referees which would have been impressive in astronomy circles. This included the Secretary General of the International Geodesy Union,

Professor Georges Perrier (Paris), Richard Schumann (Vienna) and H. Kimura (Japan), winner of the Gold Medal of the Astronomical Union. The Director of the Turin Observatory Luigi Volta, who was responsible for the discovery of several asteroids and was the great-grandson of Alessandro Volta, the inventor of the electric battery, also wrote a detailed and positive reference for Bemporad.[224]

In his application, Bemporad stated, "I have a great experience in numerical calculus and a complete knowledge of exploiting calculating machines. Consequently, I should be able to be employed in work concerning statistical researches or as a computer in insurance institutes or similar ones." He also said that he would like employment "as close as possible to Italy, or USA or Palestine."[224] It is unfortunate that information of his unique experience in computation was not forwarded to the mathematicians who established the group at Bletchley Park who pioneered computational methods for decoding secret messages. However, by this time, in 1939, it was becoming very difficult indeed for the SPSL to assist applicants and nothing transpired. Tess Simpson wrote back on 10 January 1939 with the usual response for that time: "We will do our best to help you find a suitable position, though conditions are not very good just now."[224] Bemporad's particular contribution was in accurate determinations of latitude variations which, although a major area in research in the USA and Japan, was not a significant academic subject in the astronomy research groups in the UK at that time.

In Italy, Bemporad had been made a trustee of the *Comasebit a Ventimiglia*, an Italian organisation set up to assist Jewish people, but this was shut down by the Fascist Government in August 1939.[216] Ports such as Genoa, near Turin, had become the departure point for many Jewish people fleeing the pogroms in the 1920s and 1930s in Europe and Russia to travel by boat to the USA or South America. A Delegation for Assistance to Emigrants (Delasem) was established in Genoa and Bemporad was made a member of the Turin delegation.

After Italy entered the war on 10 June 1940, many Jewish people were arrested and sent to internment camps. The Delasem then expanded its operations to assist people in the camps. Furthermore, when the USA entered the war at the end of 1941, the possibility of using Genoa as a port for departure from Italy was closed. During this period, Bemporad managed to continue with some astronomical correspondence.[216] However, Turin was badly bombed in November 1942 and many documents were lost. After the deposition of Mussolini on 25 July 1943, following the Allied invasion of Sicily and the subsequent occupation of Northern Italy by German forces, the Delasem was terminated and Bemporad and colleagues went into hiding in the Italian countryside. The new regime announced that all Jewish people should be sent to concentration camps and their belongings confiscated.[217]

Bemporad moved to Rome, which was liberated on 4 June 1944. He was then made Director of the Palestine Central Office of the Italian Zionist Organisation. This organisation had the aim of assisting the immigration to Palestine of Italian Jews who had survived. There are letters which show that Bemporad was hoping to return to Turin but he was suffering from lung cancer and died suddenly and unexpectedly on 9 July 1945.[216] He was buried in Rome and there is a fine monument at the Verano cemetery in his name. His obituary was written in the *Memoirs of the Italian Astronomical Society* by the Director of the Turin Astronomical Observatory, Professor Luigi Volta, who had been one of Bemporad's referees in his application to the SPSL.[225]

3.5 Karl-Heinrich Riewe, Part I

After an unsuccessful application to the AAC, Karl-Heinrich Riewe remained in Germany and undertook research in Berlin during the war on a project to purify uranium. At the end of the war, he was taken by the Soviets to work on their nuclear weapons programme. After being sent to a camp in Siberia, he returned eventually to West Germany and became managing director of the German Physical Society.

Karl-Heinrich Riewe was born on 26 June 1907 in Frankfurt an der Oder, Germany, to a Jewish family. He studied physics at the Technische Hochschule Berlin from 1928–1934 and attended lectures from many of the great names then based in Berlin including von Laue, Schrödinger, Hertz and Pringsheim. He also studied under Franck and Courant in Göttingen. In 1931, he worked on valve

Fig. 3.11   Karl-Heinrich Riewe, photograph used in his AAC application (1935).

devices for the Osram company, which was to be a very helpful experience for him a few years later. He obtained his Dr. phil. in May 1935 with a thesis on dielectric constants and conductivity of electrolytes. Riewe was versatile and even published with the Nobel Laureate Max von Laue on crystallography.[226]

On 8 July 1935, Riewe applied to the AAC (see Figure 3.11).[227] He stated, "As a non-Aryan, I have no prospect of getting a job in Germany, either as an assistant at a scientific institute or anywhere in industry." In his reference for Riewe to the AAC, von Laue said:

> Dr. Riewe, who is just doing his doctorate in physics at the University of Berlin, will contact the Academic Assistance Council shortly; he is non-Aryan by birth and therefore cannot find a job here. I know him personally and will be happy if he manages to get a job abroad.[227(t)]

As is described in the following discussion, a position in another country did not prove possible for Riewe until the end of the war when he went to the Soviet Union. On his application form, he stated that he had taken up the Protestant faith. The SPSL sent him information on a position at Wesleyan College in the USA. Following a suggestion from

Ralph Fowler (Cambridge University), he then wrote again to the SPSL on 6 August 1937, with an updated curriculum vitae stating that he had been able to survive in Germany "by making physical calculations and other work."[227] He also stated that he had met Gustav Wikkenhauser, a Hungarian television pioneer, who was working for a company, Scophony Ltd., based in London and had the prospect of employing Riewe providing he obtained a permit from the Ministry of Labour. Riewe asked the SPSL if it could obtain such a permit. Walter Adams, the General Secretary of the SPSL, then replied on 12 August 1937 stating that the application needed to be made by the employer and sent a suitable form for this purpose. He asked if Riewe could inform him if such an application was made. Riewe, however, remained in Berlin working for the Osram Electrical Lighting company and also published several papers on the dissociations of small molecules in journals such as the *Annalen der Physik*.

The next information the SPSL received on Riewe was a communication, sent by Wolfgang Ehrenberg, of an abstract of a paper on electron beam physics published in 1941 by Riewe and Fritz Houtermans.[228] The paper included the address of the authors as the Laboratory of Manfred von Ardenne, Berlin. This surprised the SPSL as it had a large file on Houtermans.[229] There is now an interruption in our discussion in order to describe the extraordinary life of Fritz Houtermans before continuing the story of Riewe.

3.6 Friedrich Houtermans

Friedrich (Fritz) Houtermans (1903–1966) was unique — a physicist who survived World War II despite being imprisoned and tortured by both the Nazis and the Soviets. While working in Berlin, Fritz Houtermans was an author of the first paper that suggested quantum tunnelling is required to explain the production of elements in stars. After a brief spell in industry in the UK, he moved to the Ukraine Physical Technical Institute. In the purges in the Soviet Union, he was arrested before being returned to the Nazis in a prisoner swap (see Figure 3.12). He was then released in Germany during the war to conduct research in their

Fig. 3.12   Fritz Houtermans in a Soviet prison in 1938.

nuclear programme. After the war, he eventually settled in Berne where he worked on geophysics problems.[230,231]

Houtermans was born on 22 January 1903 at Zoppot, near Danzig, on the Baltic Sea, to a Jewish family. His father was a banker and his mother was the first woman to get a doctorate in chemistry at the University of Vienna. He studied first at the University of Vienna and then in Göttingen where he received his doctor degree in 1927 supervised by James Franck.

Houtermans then moved to the Technische Hochschule in Berlin where he worked with Gustav Hertz who was awarded the Nobel Prize for Physics in 1925 together with Franck for their work on electron–atom scattering. During this period, Houtermans published some significant papers including one of the first, with Gamow, on the importance of quantum tunnelling for explaining radioactivity.[232] Houtermans then used this idea with the British physicist Robert Atkinson in the very first paper suggesting this was a mechanism that explained the production of elements in stars.[233] This was a founding paper in the field of nuclear astrophysics in which they coined the word "thermonuclear." Their idea inspired Cockcroft and Walton in Cambridge to develop an experiment which involved splitting nuclei with energetic protons and led to a Nobel Prize. The ideas of Houtermans and Atkinson were taken forward by several astrophysicists including Hans Bethe who explained the nuclear reactions occurring in the sun and other stars.[14] Bethe, who won the Nobel Prize for Physics in 1967 for this work, was himself a German Jewish scientific refugee who had originally studied under Sommerfeld and Fermi. He been helped by the AAC in the 1930s to come to Manchester and had eventually found his way to Cornell University.[14,234]

Following a major conference in Odessa in 1930, Houtermans married Charlotte Riefenstahl in a Black Sea resort. She had studied with James Franck and had been, for a short period, the girlfriend of J. Robert Oppenheimer who she had met in Göttingen and had also visited in the USA. Wolfgang Pauli, a close friend of both Charlotte and Fritz Houtermans, was a witness at their wedding, as was Rudolf Peierls. A daughter was born to the Houtermans couple in 1932 and a son in 1935 (see Figure 3.13).[230]

With the imposition of the new laws in Germany in 1933, Houtermans was under particular threat, not only because of his Jewish background but also because he had been a member of the now-banned

Fig. 3.13 Fritz and Charlotte Houtermans in 1932.

German Communist party. He then made the unusual move, with the help of the AAC, to the laboratory of Electrical and Musical Instruments Limited ("His Master's Voice"), in Hayes Middlesex, UK, in 1933.[229] There, he was allowed to experiment on Einstein's proposal that a light beam can be amplified when it is passed through a gas of electronically excited atoms. If a transformer had not burnt out, Houtermans might have achieved the first demonstration of the laser.[231]

Houtermans, having a sharp wit, was not very complimentary of the English way of life. He often met up with Otto Frisch, also a physics refugee in England, who wrote that Houtermans stated, "The British were a poor nation who lived on the residues of wool manufacturers."[235] Houtermans also complained that English restaurants always seemed to have the aroma of stale mutton and the British only had one way of cooking potatoes. Frisch, however, said that he and Houtermans both liked the Woolworths store where they could buy almost anything very cheaply, including a single sock for a friend who only had one leg, and also materials needed for their physics experiments.[235] In due course, Houtermans became famous for his quotations. He was responsible for giving the name "Martians'" to the brilliant scientists who left Hungary, Theodore von Kármán, George de Hevesy, Michael Polanyi, Leo Szilárd, Eugene Wigner, John von Neumann and Edward Teller, stating, "They are Martians who are afraid that their accents will give them away, so they masquerade as Hungarians, i.e. people unable to speak any language but Hungarian without an accent."[236]

Houtermans was also finding it very dull living in the small town of Hayes in Middlesex which had nothing like the cultural and intellectual excitement of Berlin. In the words of Charlotte Houtermans, "with ugly little houses, semi-detached, with identical gardens in streets with pretentious names."[231] They did, however, enjoy the parties thrown by Patrick Blackett in London. With his experiments in England not being successful, Houtermans looked for a seemingly safe but more scientifically exciting environment.[231]

While in England, Houtermans met Alexandr Leipunsky from Kharkov in the USSR, who had a research fellowship with Rutherford in Cambridge. Leipunsky arranged an offer for a professorship for

Houtermans in nuclear physics working at the Ukraine Physical Technical Institute (UPTI) in Kharkov, starting in December 1934. Despite intense warnings from his friend Pauli about the political risk, Houtermans decided to move to Kharkov.[231] The UPTI had been founded in 1928 by Abram Ioffe and had become the leading institute for physics research in the USSR. The future Nobel Laureate Lev Landau had been appointed head of the Theory Department there in 1932. Alexander Weissberg was a physicist from Austria who had joined the UPTI in 1931 and started the *Soviet Journal of Physics*. He, together with the UPTI Director Ivan Obreimov, realised that, following Hitler's coming to power, there was an opportunity to recruit leading German physicists to help start a new laboratory in low-temperature physics. Weissberg was assisted in his vision by Nicolai Bukharin who was one of the original Soviet revolutionary leaders. This initiative was analogous to that taken by Frederick Lindemann in the UK, who recruited Francis Simon and several of his co-workers from Breslau in 1933 to start a new laboratory on low-temperature physics at Oxford.[4]

At the UPTI, Houtermans appreciated the opportunity to interact closely with some of the greatest Russian physicists including Landau. Houtermans spoke Russian "badly but fluently with characteristic energy."[236] He published in the *Soviet Journal of Physics* with Leipunsky and Rusinow on the absorption of neutrons by silver, cadmium and boron.[237] However, by 1937, the political atmosphere in the Soviet Union was becoming poisonous and Stalin had Bukharin arrested. This, together with the recruitment of German scientists such as Houtermans, placed the physicists at the UPTI under suspicion.[231]

Houtermans himself subsequently listed the names of Schubnikov, Gorski, Rosenkiewicz, Obreimov, Fomin and Weissberg as physicists who were arrested from the institute in Kharkov.[229] The first three of these were executed in 1937 while Fomin took his own life when he heard he was to be arrested for association with Houtermans.[231] In addition, the internationally famous Soviet physicists Vladimir Fock and Lev Landau were both arrested but were subsequently released after pleas from Pyotr Kapitsa. Some scientists who had come to the USSR from Germany, including the mathematician Fritz Noether and the quantum chemist

Hans Hellmann, were imprisoned and this was noted by Houtermans.[229] Hellmann had moved from Hannover to the Karpov Institute of Physical Chemistry in Moscow. Hellmann derived an important theorem that enables electrostatic forces in molecules to be calculated from Schrödinger's wave mechanics.[238] Richard Feynman then published a similar work in 1939 on "Forces in Molecules".[239] Their contribution to wave mechanics is now known as the Hellmann–Feynman theorem. Less achieving colleagues in Moscow were envious of Hellmann's work and he was arrested by the secret police, forced to sign a "confession" and then shot.[240] This was almost the fate of Houtermans when suspicion fell on several emigrant scientists in the Soviet Union of being German spies. We have already seen how Duschinsky and Wasser were forced from their posts in Leningrad and they were perhaps fortunate at that time in not receiving stricter penalties.

On 3 January 1938, Charlotte Houtermans sent an alarming and detailed letter to Patrick Blackett in London describing recent events of her husband.[229] Blackett had worked alongside both Charlotte and Fritz Houtermans in the laboratory of James Franck in Göttingen in 1924. She stated that her husband was dismissed from his post at Kharkov on 15 September 1937. He then applied for a visa to exit Russia and went to Moscow to sort out his affairs. However, this took some time and the delay was crucial as Houtermans was arrested in the customs office in Moscow on 1 December 1937. Charlotte asked Blackett if the SPSL could do anything to help her husband. Blackett was already aware of the threat to Houtermans and had made an emergency application to the SPSL in October 1937 for a grant of £250 that he hoped would enable Houtermans to leave Russia and come back to work in the UK. This had been granted on 16 October.[229]

There followed then a number of further urgent letters to the SPSL to try to get Houtermans out of prison and find a safe haven for his family. As an example, on 24 December 1937, the mathematician Harald Bohr in Copenhagen, the brother of Niels Bohr, wrote:

> A few days ago, we got a wire from her (Mrs. Houtermans) from Riga
> that she with her two small children most hurriedly had left Russia,

while her husband was not able to leave, and that she was in Riga without any means and only possessed a German passport valid for a few days and just for returning to Germany via Tilsit. From the telegram we understood that Mrs. Houtermans was extremely excited over the possibility of being forced to go to Germany and that she — and we think with full right — expected, as the whole situation lies, to come into great danger when entering. We succeeded in getting her a visa for coming to Denmark, but it was not possible to get a visa for more than twenty days, and we have to see that she can go to some other country in the course of this time, as it seems as if the Danish authorities under no circumstances will prolong the visa (such that there is the great danger that she may be sent to Germany).[229]

Harald Bohr then wrote again to the SPSL on 3 January 1938:

After having discussed with my brother and others interested in the case, we have some days ago sent in a formal application to the Ministry of Law for an identification certificate for Mrs. Houtermans. In order to persuade the Danish authorities to give such a certificate it would however, as I understand, be of very great importance, if you could succeed in getting such assurance from the Home Office which could authorize you to send an official letter (perhaps best addressed to my brother) to show to the Danish police, as indicated in your letter, and which also could give information what travel document is claimed by the Home Office in order to admit Mrs. Houtermans to England.

As regards Prof. Houtermans himself it is just the dreadful situation that nobody including Mrs. Houtermans knows anything more than that she and my brother have written to Prof. Blackett about his position. However, we hope that there is no all too great danger for him.[229]

Charlotte Houtermans and her two children had managed to escape from Russia to Denmark from Riga. They then went on to England and eventually the USA (where she had visited J. Robert Oppenheimer before). She orchestrated a set of letters from highly influential and eminent people to save her husband. This included letters from Einstein and Blackett to the Soviet Ambassadors in Washington and London, respectively. The Soviet-supporting biophysicist J. D. Bernal from the

Cavendish Laboratory in Cambridge arranged a meeting with the Soviet Ambassador Maisky, who he knew, to discuss the matter. The British Foreign Office also suggested confidentially to the SPSL that the influential left-wing writer H. G. Wells should also do the same.[229] In due course, the Nobel Prize winners from Paris, Irène and Frédéric Joliot-Curie, together with Jean Perrin, sent a telegram in French to the Russian General Prosecutor Vyshinsky and to Stalin himself on 21 June 1938:

> I demand an explanation on the fate of eminent physicists Alexander Weissberg arrested in Kharkov on 1 March 1937 and Friedrich Houtermans arrested in Moscow on 1 December 1937. Their imprisonment threatens our political campaign against the enemies of the USSR and at the same time is incomprehensible for friends of the USSR who are convinced that Weissberg and Houtermans are incapable of any hostile act against socialist construction and we are convinced their arrest signifies a serious error. We request your special attention for their case which highlights political importance and request an urgent response.[229(t)]

Niels Bohr had written to the SPSL to say he was not writing letters to the Soviet leaders as he was concerned it could lead to the arrest of Landau and Kapitsa whose precarious position he heard about during a visit to Russia in the summer of 1937. However, he acknowledged that Frédéric Joliot-Curie in Paris had closer links with the Russian authorities. None of these efforts, however, led to the release of Houtermans from prison in Russia. It is possible, however, that the concerns made on Houtermans's behalf by so many highly prominent people demonstrated his importance and may have influenced the Soviet authorities to keep him alive. On 14 June 1938, Leipunsky was arrested and, following torture, made to sign a confession stating that Houtermans was a German spy.[230]

The efforts to bring Charlotte Houtermans to London succeeded but she was becoming increasingly frustrated and concerned. On 11 February 1939, she sent a telegram to the infamous Laurentia Beria, Head of the Soviet NKVD:

As wife of Dr. Fritz Houtermans, physicist from the Ukrainian Physico Technical Institute Kharkov, I am appealing to your generosity of letting me know what has happened to my husband. I and my children, the younger being born in Kharkov, were separated from him on 1 December 1937 when we were in Moscow on the eve of our departure from USSR, all holding valid visas for exit. My husband was arrested on December first 1937 in Moscow at the custom house …

I assume that he was taken to Kharkov and later Kiev but his correct address was never given to me nor has any statement as to the charges against him ever reached me. I am very anxious about his fate. My husband is very well known in scientific circles in all countries. When asked about him as frequently happens I am completely at a loss of giving a satisfactory explanation of his disappearance. My husband and I have been grateful for the hospitality which was offered to us by the USSR and on which my husband's scientific work especially greatly benefitted. Do please give me some information about him and reassurance as to his state of health. I am confident that his case will be justly dealt with and I shall be very thankful for every effort made to ensure his release for which I, my children and his aged mother are longing day to day.[229]

In April 1939, Charlotte Houtermans moved with her children to the USA where their grandmother already lived. She was appointed to a position first at Vassar College, where she had taught in 1930, and then at Sarah Lawrence College. She managed to take up research again and published on the ion bombardment of metals.[241] She also wrote to Eleanor Roosevelt about her husband's plight and received several letters in response.[242]

Fritz Houtermans's subsequent writings show that his experiences in prison were appalling. He was moved between prisons in Moscow, Kiev and Kharkov with often over 100 prisoners confined to a small cell. He was sometimes interrogated for nights on end by the NKVD with no sleep. He was ordered to sign statements against several of his colleagues including Landau and Weissberg.[231] Weissberg subsequently published a widely read book which describes in vivid detail the Soviet terror.[243] Finally, after he was threatened that his children would be sent to an orphanage, Houtermans confessed to being a German spy.[230]

Then, following the Molotov–Ribbentrop pact of August 1939, and with Germany and the Soviet Union still not at war, Houtermans was handed over to the Gestapo in a prisoner swap on 25 April 1940.[231] It is quite possible he had been kept in a Soviet prison for over two years with this purpose in mind. Eleanor Roosevelt was informed of Houtermans being sent back to Germany on 26 June 1940 by the US Ambassador to the USSR, Laurence Steinhardt, and she at once informed Charlotte Houtermans.[242] Weissberg was also swapped in a prisoner exchange and then had a remarkable series of escapes from the Nazis in Poland where he had been sent.

Being both Jewish and a member of the banned German Communist party, the position of Fritz Houtermans in Germany was extremely perilous and he was interrogated by the Gestapo. However, the highly influential and persuasive Max von Laue and Walter Gerlach managed to convince the Nazi authorities that it would be advantageous to them to allow Houtermans to be freed from prison to work in the private laboratory of the physicist and inventor Manfred von Ardenne (1907–1997) in Berlin.[231] This was where Houtermans collaborated and published with Riewe.[228] Both Riewe and Houtermans were "non-Aryans" but working with von Ardenne appeared to give them protection.

Von Ardenne had many electronic inventions to his name including the scanning electron microscope and one of the first television systems made from cathode ray tubes, and this gave him unusual freedom for research and recruitment of scientists, including "non-Aryans" and political dissidents, in Nazi Germany.[244] Von Ardenne and Houtermans even commissioned Schrödinger's close friend Hans Thirring, who had been dismissed from his chair at the University of Vienna after the Anschluss because of his association with anti-Nazis, to perform calculations on the power consumption of a cyclotron. Thirring visited Houtermans and von Ardenne in war-time Berlin and subsequently sent Houtermans, on 15 July 1942, his completed manuscript on calculations for a "Berlin cyclotron" that was to be constructed in the laboratory of von Ardenne and based on a design developed in 1936 by Eugene Wigner at Princeton.[245]

From von Ardenne's laboratory, Houtermans also authored a classified paper on the reaction of uranium-238 with neutrons to produce a new, highly radioactive element plutonium and further neutrons.[246] This mechanism produced a chain reaction that had the potential for use as a nuclear weapon. The new element plutonium did not require the difficult physical separation of isotopes of nearly the same mass as was the case for the 235 and 238 isotopes of uranium. Houtermans's paper was seen by Heisenberg and others who were directing the German nuclear research programme. However, it seems they concluded that the technical difficulties of producing a critical mass of plutonium were too complicated to overcome on a short time scale.[247]

It is ironic that Enrico Fermi in the USA did surpass these challenges. His time in Göttingen had overlapped with that of Houtermans, and he had gone to the USA in 1938 straight after receiving his Nobel Prize for Physics in Stockholm following the imposition of new anti-Semitic laws in Italy. He did not return again to Italy and pioneered in Columbia University and Chicago the first artificial nuclear reactor. This was scaled up at Oak Ridge and Hanford to produce the plutonium at the core of the first nuclear explosion at the Trinity Site in New Mexico in 1945.[218] Eugene Wigner subsequently reported that during the war he had received an anonymous letter from Geneva with the phrase "Hurry up, we are on the track." It has been widely stated that this message was sent by the famously mischievous Houtermans and that it had a significant influence on accelerating the nuclear Manhattan project in the USA.[230,248]

While working in the von Ardenne laboratory, Houtermans was even allowed to visit and make a report on his old laboratory in Kharkov in 1941 which, by then, was occupied by the Germans. When he was in prison in the Soviet Union, Houtermans shared a cell with a Professor of History from Kiev, Konstantin Shtepa. During his visit to Kiev in 1941, Houtermans met up again with Shtepa and let him have some of his food rations. They continued to correspond during the war and Houtermans arranged for more food to be sent to his friend. He even helped to prevent Shtepa's son from being sent to a concentration camp.[231] Houtermans subsequently claimed that this visit was made to assist his former colleagues but some scientists in the Soviet Union did not see things that way.[229]

Near the end of the war, Houtermans, with the help of Heisenberg and Gerlach, who had influence with the Nazis, was appointed to a position in Göttingen which later became part of the Allies' sector of Germany. He was fortunate that this was not in the Soviet sector as he may well have been arrested again. An alarmed Max Born wrote to the SPSL on 10 November 1947:

> About two years ago I had correspondence with you regarding Professor Houtermans. When I was in Moscow in 1945, Professor Kapitsa called him a traitor, and said that the Russians would hang him if they found him. He said that Houtermans had come with the German Army to Kharkov where he previously had been Professor, and had denounced his former colleagues to the Gestapo. Although I did not believe the whole story, I thought it better to warn Houtermans about what the Russians had told me, in case he should be caught going into the Russian Zone. Later I heard of his strange behaviour with regard to his wife, who had done everything possible to him when he was in the Russian prison, and who now lives a miserable existence in America. Houtermans did not join her there, in fact I learned that he had married another woman.
>
> Some months ago, a young Austrian Dr. B. Touschek, who is now working under Professor Dee in Glasgow University, came to Edinburgh, and said that he was a friend of Houtermans, so I told him some of what I had heard of Houtermans. Touschek communicated with Houtermans, and told him what I had reported, and he has had a reply from Houtermans in which he defends himself. I enclose a copy of the relative parts of Houtermans' letter to Touschek. I beg you to treat it as completely confidential, but to form your own opinion. I shall send a copy also to Professor Blackett, who was interested in this matter. Concerning my own opinion, I think that Houtermans' explanation sounds quite credible.[229]

In his letter, Born also made a comment on "the Russians" who he had previously recommended for research positions: "My trust in the Russians (and I suppose that of many others) has been severely shaken during the past two years. They seem to be little better than the Nazis. Houtermans's letter contains a lot of information on what they did to a number of Russian physicists."[229] Ursell replied to Born: "Professor Houtermans explanation on the whole confirms earlier reports which had reached us, and I hope the Russians will not take further action ... On a more cheerful note, I have now the pleasure of working on the International Club House Committee with your daughter Mrs. Newton-John."[229] Here, Mrs. Newton-John is Irene, Max Born's daughter who married Brindley Newton-John. He had worked on German intelligence in the Second World War at the Bletchley decoding centre and eventually immigrated to Australia where he was Professor of German at the University of Melbourne. His daughter was Olivia Newton-John, the famous singer and actor.

The letter from Houtermans, which was forwarded to the SPSL by Born, defended his position on his 1941 visit to Kharkov and Kiev:

The fact that my behaviour towards Russia is being criticized seems so grotesque to me that I could not answer at all if the reproach did not come from Born, whom I see and love so much ... I was arrested because of my origins as a German, among German theorists I think of Hellmann and Noether (there is no news of the former, I met the latter in a Butyrka cell in early 1940, he had been sentenced to 25 years in prison for "spying" whose confession had been tortured to obtain). As for my experiences ... I was handed over into the hands of the Gestapo, to the very country in which I had been politically persecuted ... In a commission of three scientists I was then asked to undertake a journey of a few days each to Kiev and Kharkov to check the conditions of the institutes there. I accepted the invitation because its rejection in my situation was impossible and because it gave me a unique opportunity to saving and rescuing institutes and people. Surely the Russians cannot reproach me if they have extradited me to Germany for political reasons, after I had refused the intervention of the German consulate despite my three-year imprisonment.

In fact, I managed to save a few institutes and arranged for the supply of rations to the workers (the Germans in the big cities gave no rations at all to the civilian population) ... I also worked to save Leipunsky and Landau whose capture had been reported in Berlin (declaring that Landau was a Caucasian of German descent and Leipunsky had been persecuted by the Russian government) ... Needless to say, I still value Russian science and its proponents, despite all that has been said. I studied there, and feel connected to them and hope that my return will be possible when there is communication free from all political constraints.

What has become of Schubnikov (one of the most important Kältephysikers), Gorski, Obreimov, Rosenkiewicz, Fomin, Weissberg, to name but a few employees of the UPTI? What has become of Hellmann and Noether? Can't Born do something for them? Can't he try to do something for Richard Becker's son, who has been a prisoner of war since the taking of Lvov. Is something known about the destiny of the Russian astronomer Gerasimowitsch etc. who all fell victim to the NKVD? What has become of Krutkov, Bursian and Frederiks. Is Fock free again? When the Academic Assistance Council made enquiries about the German victims what about the Russian ones?[229(t)]

An additional self-explanatory letter defending Houtermans had been sent on 22 November 1947 to the SPSL by Marie Rausch von Traubenberg who had herself escaped from Germany and was living in Swansea:

I want to certify the following facts about Professor Fritz Houtermans:
My late husband, Heinrich Rausch von Traubenberg, Professor of Physics, and I met Professor Houtermans, an old friend of ours, just after his return from Russian prisons in the year 1940. He then had exchanged with other prisoners and thus came to Berlin. We were very good friends with him and saw him almost every day during those years of war until we were bombed in February 1944 and left Berlin. So I am able with absolute certainty to certify that Houtermans always was a strict opponent of the Nazis, not only by words but by many deeds. Thus he took the job to go to Kiev and Kharkov with the Germans with the direct purpose to try and save his former colleagues and to hinder the destruction of the institutes. Indeed, I know that he was very glad to have been able to prevent a lot of evil and to help his Russian friends. Though himself in a very bad state of undernourishment he often managed to send his own rations to them. Nevertheless it could very easily happen that his coming together with the Nazis is misunderstood, because the other side might see only what actually happened and not what Houtermans succeeded in preventing, the more so as he had to camouflage very cautiously.

During this journey to Russia Houtermans made the acquaintance of the Obersturmbannführer der SS Theodor Cammann, with whose help he was able to save the life of many persons, for instance my own life and that of my mother. Furthermore he saved Richard Gans, Professor of Theoretical Physics, who had to carry stones for ten hours a day and, being an old man, would have died very soon. He saved him, too, from deportation.

Then I know that very often he hid Jewish persons, who he scarcely knew, in his lodgings. All this was done with very great person risk, so that my husband and I often were much afraid the Nazis might catch him. In February 1944 we left Berlin and did not see Houtermans for half a year. Then on September 19th I was arrested, and my husband a few minutes later died from heart failure, he at once with great risk came to the funeral and wanted to help me to hide. Although this was not possible and I had to go back to prison and later on to Theresienstadt, it was owing to him and Cammann's help, together with applications made by Professor Otto Hahn and others that there was great delay in my being sent to Theresienstadt. And it actually was this delay that saved my life, otherwise I almost certainly would have been sent to the gas chambers.

Therefore, I have the deepest desire to help Professor Houtermans by this statement as he helped me. As to his confession made during his imprisonment in Russia to be a spy of the Gestapo, I know with certainty that it was false and only made under great mental stress and fear for the lives of his children. But I think that this confession was annulled by himself before he left Russia. My husband, whose character and adverse feeling against the Nazi regime and everything unjust is well known, I think, often discussed these matters with Houtermans and had a high opinion of his ethical qualities.[229]

The SPSL received a letter from Michael Polanyi sent on 2 February 1948 from Manchester University asking if the SPSL could give assistance in obtaining a permit for Houtermans to come to lecture in Manchester. Polanyi, described by Houtermans as one of the "Hungarian Martians," explained that he was an old friend.[229] He was clearly aware that it was not a straightforward process to obtain permission from the British authorities for scientists who had worked in Germany during the war to visit the UK. Indeed, there had been a similar application for Heisenberg to visit the UK at this time which, although successful, had its complications.[4] Tess Simpson herself helped with the arrangements and Houtermans sent her a warm letter of thanks afterwards. The brother of Ilse Ursell, the applied mathematician Fritz Ursell, was then researching in Manchester and Ilse wrote to Houtermans to say she also was particularly pleased to assist with his visit.

After the war, Houtermans continued to research in Göttingen, publishing on radioactivity and the properties of neutrons. However, the funding and facilities for research in post-war Germany were limited and German scientists also had significant restrictions on travel. Houtermans developed a new interest in the science of meteorites and in 1952 moved to the University of Berne in Switzerland to set up a new laboratory in the geophysics field.[231]

He was married four times. Charlotte, who worked so hard to save his life in the Soviet Union, was his first and third wife. They divorced in 1943 when she was in the USA and he in Germany. After the war, they met up and were married again in Berne in August 1953 with Pauli, who loved symmetry, once more being a witness.[190] Charlotte had known Pauli since the 1920s and they had corresponded regularly.[242] However, Houtermans and Charlotte separated again after only a few months and their marriage ended for a second time. Houtermans then married Lore Müller, who was the sister of his step-brother's wife. Pauli sent a telegram that stated, typically and simply, "the usual congratulations".[190] The larger-than-life physicist Fritz Houtermans died at the age of 63 on 1 March 1966 from a stroke following lung cancer. He had always been a heavy smoker. Charlotte, a hugely courageous part of his story, lived till the age of 93 and died in Minnesota in 1993.

3.7 Karl-Heinrich Riewe, Part II

It is quite possible that Houtermans and Riewe published their joint paper in 1941 to inform their scientific colleagues around the world that they were still alive.[228] A 1957 report from the CIA, which was declassified in 2018, enables one to follow Riewe's movements for the years after

1941.[249] During the war, and after working in von Ardenne's laboratory, Riewe took up a position at the industrial laboratory of Nikolaus Riehl in Berlin run by the company Auergesellschaft. Here, there was a major effort in the production of high-purity uranium oxide and Riewe's expertise in high-frequency spectroscopy was put to use. Riehl himself had previously undertaken research with both Lise Meitner and Otto Hahn.

At the end of the war, the Soviet authorities made it a priority to take into custody any scientists who might be useful for them and especially those linked to German nuclear research. At this time, Houtermans was fortunate to be placed in the western sector in Göttingen, but Riewe was taken to Moscow to work on the Soviet nuclear weapons programme, together with 10 other German scientists headed by Riehl.[250,251] Riewe's patron in Germany, Manfred von Ardenne, also joined the Soviet nuclear research programme, as did his former teacher in Berlin and Nobel Laureate Gustav Hertz.[251] Riewe's family were allowed to go with him to Russia, although he subsequently divorced his wife.

Riewe was placed in charge of commissioning the very latest electrical devices for the laboratory at Factory 12 Elektrostal, 58 km east of Moscow.[249] Substantial funds were provided for this purpose through Beria, Stalin's Deputy Premier. Riewe even managed to commission some of the most expensive equipment from the USA, as officials were not then aware of its purpose. The Elektrostal Laboratory was successful in producing uranium metal of very high purity which was required for the Soviet nuclear programme, and Riehl was awarded the Stalin and Lenin Prizes.

In February 1947, Riewe was transferred to a laboratory in Obninsk, 110 km southwest of Moscow, under the direction of Heinz Pose, who had also worked previously in the German nuclear research programme. Riewe had originally signed a two-year contract and he went on strike in 1948 with the hope of being able to return to Germany.[251] This was not acceptable to the Soviet authorities as the top-secret research was vital to their efforts to produce nuclear weapons. Riewe was given a 25-year sentence and sent to the Soviet Gulag internment camp in the city of Dzhezkazgan, Kazakhstan.[251]

A report states that Riewe then completely disappeared and there was further information that he was executed.[251] However, this was not the case and, after the death of Stalin, Riewe was allowed to leave the USSR in 1955. He then managed to make his way to West Germany where he lived in Hanau and worked in the patent department of W.C. Heraeus. The British Foreign Office had been keeping a special watch on German scientists who were returning from the USSR and Riewe was interviewed about life in the camps of the Gulag.[252,253] He described a major uprising in the Kengir camp which was also featured in Solzhenitsyn's *The Gulag Archipelago*.[254]

In 1961, Riewe was made secretary of the German Physical Society (DPG) and was managing director from 1962–1972. He is credited with helping to rebuild and unify the DPG.[255] Max von Laue had been president of the DPG from 1931–1933 and had been active in its reformation after the war. Riewe's published work with von Laue would have helped to provide the respect needed to take on his administrative role with the DPG. It seems, however, that Riewe's work in Nazi Germany on uranium purification and on the post-war Soviet Nuclear Weapons programme was kept quiet and was not mentioned in subsequent reports on his life.[255,256]

CHAPTER FOUR

Top-Secret Refugees

A small number of the scientific refugees who were experts in physics came to the UK and then undertook highly confidential nuclear research during the war for the governments of the UK and USA in secret locations. The SPSL lost communication with some of those physicists who had moved temporarily to the USA. After the war, these individuals received letters from the SPSL asking about their whereabouts. The responses were illuminating.

4.1 Otto Frisch

Otto Frisch (1904–1979) sent a very understated reply to the SPSL on 27 November 1945 from P.O. Box 1663 in Sante Fe, New Mexico, on the first realisation of the production of the atomic bomb:

> Early in 1939, I had published, partly together with Lise Meitner, some short papers which helped to establish the existence of atomic fission. When the essential possibility of nuclear chain reactions became evident I gave some thought to the technical realisation of such processes and early in 1940 wrote a report to the Government, together with R. Peierls, pointing out possibilities of making an atomic bomb. (This is mentioned in the White Paper on the atomic bomb). Since summer 1940 I have been employed by the Government working on the development of the atomic bomb. Most of the time I worked at the University of Liverpool, under Sir James Chadwick; for a few months I worked in Oxford at the Clarendon Laboratory. The details of my work cannot of course be described at present.

Fig. 4.1   Los Alamos badge photograph of Otto Frisch, 1943.

In November 1943 I became a British Subject and was immediately sent to the United States, where I have been working as a member of the Ministry of Supply Mission, at the big research establishment at Los Alamos, New Mexico, which had been set up to handle the numerous scientific and engineering problems connected with the actual construction of atomic bombs ... I expect to return to England in a few months' time when I will probably join the Government Establishment for Atomic Power at Harwell in a position of some responsibility.[257]

So, by the end of 1945, it was becoming clear that the work of the AAC/SPSL in helping scientific refugees move to the UK had contributed, at least indirectly, to the development of the atomic bomb. Frisch, a nephew of Lise Meitner, had joined Blackett's research group at Birkbeck College, London, in 1933. He had then gone to work with Niels Bohr in Copenhagen. In 1939, he wrote the paper on atomic fission with Lise Meitner, mentioned in his letter, in which the mechanism for the splitting of uranium atoms upon collision with neutrons was described.[258] Frisch returned to the UK in 1939 with the help of the SPSL and he then wrote his famous report with Peierls on the critical mass of uranium-235 that is needed to cause a nuclear explosion.[259]

However, following the German invasions of several countries in the spring of 1940, the British Prime Minister Winston Churchill instigated a "collar the lot" policy of interning all German nationals in the UK.[260] This even included many Jewish scientists who had escaped from the Nazis. The SPSL was then involved in a great deal of work, together with universities, the Royal Society, other institutions and prominent individuals to make the case for the release from internment of those scientists who were doing important research of national importance. The SPSL helped to make cases for over 500 individuals, several of whom had not applied to the AAC or SPSL in the first place such as Klaus Fuchs and the future Nobel Laureate Max Perutz.[3]

An alarmed Professor Mark Oliphant from the University of Birmingham was well aware of the world-changing memorandum written by his colleagues Peierls and Frisch. He was a member of the UK MAUD committee for directing secret research which had been set up in 1940. Oliphant wrote to Tess Simpson on 10 July 1940:

> I am glad to know that your Society is applying for the release of interned refugee scholars and scientists who, prior to internment, were doing work of immediate national importance. With regard to your questions concerning Dr. O.R. Frisch, I can tell you the following. Dr. Frisch has been concerned, together with Professor Peierls in putting forward some suggestion of great national importance, and work has been taken up and these possibilities are being investigated under the Air Ministry. Arrangements are being made for Dr. Frisch to work in Liverpool with Dr. Chadwick, who is responsible for this work, notwithstanding the fact that he is in a protected area. The Air Ministry at the same time have agreed that Dr. Frisch may be employed on this work. I am sure that it is only a question of time for the application to go through before he is established in Liverpool.
>
> Dr. Frisch has, of course, been in this country for far too short a time to be able to apply for naturalisation, but I can give you my strongest assurance of his integrity, and his loyalty to this country. He is an old student and colleague, and a great friend of Professor Niels Bohr, and I can say little more in his favour.[257]

Other influential colleagues of Frisch also made the case successfully that he should not be interned. He was allowed to continue his important secret work, first in the UK and then in the USA (see Figure 4.1).[235] In 1947, after working at the Harwell nuclear research establishment, Frisch was appointed to the Jacksonian Chair of Natural Philosophy at the University of Cambridge. He was the only physics refugee who obtained a Professorship at Cambridge, despite the fact that several highly distinguished refugee physicists had temporary appointments there in the 1930s.[261]

4.2 Klaus Fuchs

After the war, Ilse Ursell also wrote to Klaus Fuchs (1911–1988). He was not Jewish but had been a member of the Communist Party in Germany and had moved to the UK in 1933 to undertake a Ph.D. with Nevill Mott in Bristol. Fuchs' succinct reply to the SPSL on 12 December 1945, from the same P.O. Box 1663 in Sante Fe, New Mexico, used by Frisch, is of some interest:

> Since 1941 I have been engaged in research for the development of the atomic bomb; first on the research team of Professor Peierls in

Fig. 4.2   Klaus Fuchs after his arrest by the British Security Services in 1950.

Birmingham, and later in the United States in New York and Los Alamos. You will appreciate that at present I cannot give any more details.[262]

Fuchs was not prepared to give details of his research to the SPSL but he was very willing to pass on key atomic secrets to Russian agents. Indeed, just three months before writing this letter, Fuchs had passed on detailed notes near Sante Fe on 19 September 1945 to his contact "Raymond" (Harry Gold).[263,264] This included a description of the science and outcome of the recent Trinity nuclear test and the latest secret research in the USA including the first ideas of a hydrogen bomb. After the sensational trial of Fuchs for scientific espionage in 1950 (see Figure 4.2), there was criticism of the AAC/SPSL in the national press for freely assisting scientific refugees without sufficient vetting. An alarmed Tess Simpson, who by now had transferred to the Society for Visiting Scholars, wrote to Ilse Ursell on 2 March 1950:

Last night's "Evening News", in its report on the Fuchs trial, states that he had a grant from S.P.S.L. in 1933. To best of my recollection, we did not hear about Fuchs' existence until he had been in this country for some years; in any case, in 1933 it was the A.A.C., it only became S.P.S.L in 1937. I dare say I shall be questioned about this. No doubt Fuchs' folder is inaccessibly stored, but there should be information on the back of his card. Could you please let me have what information you can get at? It is a pity that I did not ask you for a copy of his curriculum vitae when he became a member of this Society — but how could I have known that I might need it for this? I may say that the reports in the evening papers were full of other unnecessary inaccuracies, e.g. they said that the father is a professor of physics, when he is really a professor of theology as had already been widely published.

I myself don't remember any grant having ever been given to Fuchs. I thought that our first real contact with him was when we applied for his release from internment, along with 550 other intellectuals. If you can lay hands on a copy of his curriculum vitae, or let me have a copy of our statement to the Home Office when we applied for his release from internment, I should be very grateful.[262]

The first that the SPSL had heard of Fuchs was in a letter dated 27 October 1937 from Max Born:

> I have here another young German Dr. Fuchs, who has worked some years with Prof. Mott in Bristol. He is a brilliant young fellow who has done already excellent work here in the short time of his presence. I am very interested in his future. At the moment I have succeeded to get a grant of £120 for him from the fund mentioned (at this University), and an additional sum of £20 — given by a friend who wishes not to be known. But I do not know whether I shall get this money next year again; in this case I should ask your Society for help, and I wish already now to direct your attention to this very able young man, Dr. Fuchs.[262]

Unlike Frisch, Fuchs was interned in 1940, and Max Born at once asked the SPSL for assistance in securing his release together with another of his research assistants, Walter Kellermann. Tess Simpson responded on 25 May 1940: "In approaching the Home Office we shall certainly point out that these two scientists are capable of undertaking work of immediate national importance."[262]

Fuchs was interned in the Isle of Man and then in Canada. He returned to the UK in January 1941. Born did not then have funds to support him but Peierls was able to provide a teaching appointment in Birmingham. Fuchs at once impressed Peierls with calculations on neutrons scattering from uranium. By June 1942, he had signed the Official Secrets Act and was fully involved with top-secret nuclear research "of immediate national importance."

CHAPTER FIVE

Refugees in Mathematics

It is of interest to examine whether the experiences of refugees from the fields of chemistry and physics described in the previous chapters apply to other subjects as well. The AAC and SPSL assisted refugees right across the board of academic subjects. In the 1930s, pure mathematics was well developed in several European countries and also in the UK. Applied mathematics, on the other hand, although a major subject in some German universities, had not yet developed into a separate subject in the UK.

In comparing applicants in mathematics with those in physics and chemistry, we have observed the same subtle nuances. A remark from one of the leading mathematicians in the UK, such as G. H. Hardy at Cambridge or J. H. C. Whitehead at Oxford, had a major influence on whether it was going to be possible for the AAC/SPSL to find a place for a particular scholar. Applications from distinguished and well-established mathematicians were not received nearly as enthusiastically as those from more junior and up-and-coming scholars. They were much easier to place in temporary positions with minimal financial support and with their whole career ahead of them and the potential to publish groundbreaking work.

A good example is Hans Heilbronn (1908–1975) who applied to the AAC in October 1933 at the young age of 25.[265] The reports from UK mathematicians were uniformly positive. Even the normally highly critical G. H. Hardy, from Trinity College, Cambridge, endorsed Heilbronn on 7 April 1934: "He has finished an exceptional piece of work which will make a sensation when it appears, and add greatly to his status."[265] A similar remark had come from the distinguished number theorist Louis Mordell at the University of Manchester, who said on 11 August 1933, "He is young ... and would not be difficult to provide for ... he would be the obvious type of person who ought to have a fellowship." The British universities were almost fighting over Heilbronn, and after brief stays at Bristol and Manchester, he was elected to the prestigious Bevan Fellowship at Trinity College, Cambridge. After the war, Heilbronn was appointed as Chair of Mathematics and Head of Department at Bristol University and was elected a Fellow of the Royal Society. His influence has remained significant to the present day: the Heilbronn Institute for Mathematical Research is a major national centre supporting mathematics in the UK.[266]

As was the case for chemists and physicists, the AAC and SPSL did manage to facilitate many refugee mathematicians to come to the UK.[267,268] However, the applications of a small number discussed in this book were not successful. The lives of several of these mathematicians have already been well documented and we will only summarise the salient features here.

5.1 Ludwig Berwald

First in Munich and then in the distinguished mathematics department of the German University in Prague, Ludwig Berwald (1883–1942) made many lasting contributions to differential geometry. He died tragically in the Lodz Ghetto in 1942.

Berwald was born on 8 December 1883 in Prague where his father Max owned a well-known bookshop. His family was Jewish and they moved to Munich where he studied mathematics and physics at the Ludwig Maximilian University. There, he wrote a thesis on the mathematics of surface curvature, and he also studied under Heinrich Burkhardt who had been an official examiner of Albert Einstein's thesis.[269] In the following years, Berwald gained a significant international reputation for research in differential geometry. In 1915, he married Hedwig (née Adler) from Prague and many visits there put him in contact with distinguished mathematicians in the city of his birth (see Figure 5.1). He was appointed a full professor at the German University in Prague in 1924 and became

Fig. 5.1.   Ludwig and Hedwig Berwald (née Adler), 1922.

Chairman of the Mathematics Department in 1929 when George Pick retired.

However, major changes quickly took place in Prague after the Munich Agreement of September 1938. In December 1938, the German University in Prague introduced its own anti-Semitic regulations and Berwald was dismissed from his professorship on 10 January 1939. His application to the SPSL was received on 30 May of that year.[270] He named an exceptional group of leading mathematicians as referees including Elie Cartan (Paris), Tullio Levi-Civita (Rome), Oswald Veblen (Princeton) and J. Henry C. Whitehead (Oxford), who all confirmed Berwald's reputation for major research achievements in the field of differential geometry. At the age of 56, Berwald might have seemed to be too senior to make mathematical advances, but Veblen stated, "he is still at the full tide of mathematical production." On 4 May 1939, Whitehead wrote to the SPSL:

> Berwald seemed undecided as to whether he would apply for a grant at once. Of course he had been dismissed and the future was very gloomy. But he did not seem to be in immediate and serious danger – or rather

he did not seem to fear such danger. I am not sure he had quite realised the terrible time that even such distinguished men as he are having.[270]

By 1939, it was very late to find an appointment even for someone of Berwald's distinction and productivity. On 10 August 1939, Tess Simpson wrote to one of Berwald's referees, Dirk Struik, at MIT, saying, "The market for mathematicians is overcrowded. We have already helped quite a number of mathematicians from Czechoslovakia and our advisors have warned us that it will not be possible to absorb any more in this country."[270]

In June 1939, the Office for Jewish Emigration was established in Prague by Adolf Eichmann. Like similar offices set up by Eichmann in Vienna and Berlin, the Jewish leaders of Prague were given authority to transmit his orders to the local Jewish community.[22] The confiscation of their property was introduced, and Jews were denied normal civil rights and excluded from political and social life. This initiated an exodus of Jewish people from Prague.

Even after the start of the war, Berwald continued to do original mathematical research in Prague. The *Annals of Mathematics* published his posthumous manuscript on Finsler and Cartan Geometry in 1947 which contained work he had completed just before he was arrested by the Nazi authorities.[271] In September 1941, Hitler ordered the deportation of Jewish people from Prague and this was implemented by Reinhard Heydrich, the Reichsprotektor of Bohemia and Moravia.[22] Berwald and his wife were sent to the Lodz Ghetto in Transport C on 22 October 1941. The conditions at the ghetto were very rudimentary, with 55 people to a room of 400 square feet and hardly any rations of food.[269] Berwald died in Lodz on 20 April 1942 of heart failure at the age of 59 just after Hedwig's death on 27 March 1942.[269] The SPSL file states erroneously that he had committed suicide in 1939.[270]

Ludwig Berwald's enduring contributions to differential geometry have become even more important in the present day. Berwald has given his name to the mathematics of curvature, inequality, metrics, sprays, spaces, connections and Finsler geometry.[269] He is considered one of the major mathematicians of the 20th century.

5.2 Walter Fröhlich

While working in Prague, Walter Fröhlich (1902–1942) made significant contributions to pure mathematics in a similar way to Ludwig Berwald. He was also transported to the Lodz Ghetto.

Fröhlich was born in Liboch near Prague in Czechoslovakia on 2 December 1902. His father was a medical doctor who had been mayor of that town. Although the family was of Jewish heritage, it had converted to Catholicism. Fröhlich initially studied at the Deutsche Technische Hochschule in Prague, which was particularly strong in applied mathematics. He then transferred to the more distinguished German University in Prague. His thesis on affine differential geometry was supervised by George Pick and Ludwig Berwald. Fröhlich then became Professor for Mathematics and Descriptive Geometry at the German State Real Gymnasium in Prague in 1927, and also lectured at the German University in Prague.[272] He established a reputation as a fine teacher. In 1930, he married Elise (née Goliath) who was from the town of Trutnov (see Figure 5.2).

Fig. 5.2.   Walter and Elise Fröhlich (née Goliath), 1927.

Fröhlich had published significant papers on geometry and topology with original applications to the mathematical properties of braids. Being of non-Aryan heritage, despite being a Catholic, he was fired from his positions in Prague on 1 February 1939 following severe pressure from the German government on the German-speaking universities in Prague.[272] He applied at once to the SPSL on 4 February 1939, shortly before Hitler himself arrived in Prague on 16 March.[273] This was a very late application to the SPSL, which had previously assisted several applicants from Germany and was now receiving numerous applications from mathematicians in countries that had more recently been invaded by the Nazis, such as Austria and Czechoslovakia. Fröhlich's referees were Ludwig Berwald, Paul Funk, Karl Löwner and George Pick, all distinguished mathematicians in Prague who were themselves persecuted by the Nazis. Indeed, Einstein himself had briefly been a professor in Prague in 1911 and his theory of general relativity was influenced by colleagues such as Pick and Erwin Finlay Freundlich, who was also a referee for Fröhlich.

Fröhlich's application to the SPSL was sent on to Henry Whitehead at the University of Oxford, who replied on 30 March 1939: "Some of us at Oxford are interested in his recent work and would gladly make him welcome." Tess Simpson responded to Whitehead in cautionary terms:

> Before considering a grant for Dr. Fröhlich our committee would wish to know what chances there are of Dr. Fröhlich obtaining a permanent position in a reasonable time. We have given grants recently to quite a number of mathematicians from Czechoslovakia and the committee is a little scared of having more on our hands than can eventually be absorbed.

Whitehead, who had very high standards, then replied on 6 April 1939, from his home at 22 Charlbury Road, Oxford: "He is a good mathematician but he has not yet done anything of really first-class interest ... my impression is that a man of Fröhlich's age who has published good papers on a subject of fairly wide interest (which he has) has a reasonable chance of a job in America."[273]

Following the Munich Agreement of September 1938, some British organisations were inclined to give extra help to refugees from Czechoslovakia. Accordingly, the British Committee for Refugees from Czechoslovakia (BCRC) and the Czech Refugee Trust Fund worked closely with the SPSL.[274] Arthur Beer was a distinguished astronomer and refugee from Germany who had been an associate of Einstein. He was working at the Mill Hill Observatory in London and had been named as a referee by Fröhlich. He made an urgent telephone call on 1 June 1939 about Fröhlich to David Cleghorn Thomson, General Secretary of the SPSL. Thomson then wrote to the BCRC on 1 June 1939: "We have just heard that Walter Fröhlich has had his house visited by the police" and asked if there was a special kind of visa which would enable Fröhlich to leave Prague at once.[273] Fröhlich had written directly to Edmund T. Whittaker, a leading mathematician at the University of Edinburgh, who then passed the letter to Thomas MacRobert in Glasgow, whose work on complex variable theory was closer to Fröhlich's interests. MacRobert informed the SPSL that he could provide a very small grant of £25 to Fröhlich for help with exercises and tutorial work.[273]

This financial support was sufficient for the SPSL to apply to the Home Office for a visa for Fröhlich to enter the UK. The Home Office in turn instructed the British Passport Office in Prague on 12 June 1939 to provide a visa to Fröhlich. However, the Home Office had omitted to mention the need for a visa for Fröhlich's wife, Elise. There was also a requirement to confirm Elise's full name, and the communications on this caused the delay of a month in obtaining the visa for her. Fröhlich and his wife then had to obtain passports as citizens of the German Reich and also receive permission from the Office for Jewish Immigration in Prague to leave the country, which caused a further delay.[272] However, on 3 September 1939, war had been declared between the UK and Germany and the visas for Fröhlich and his wife at once became invalid. This was explained in a subsequent letter from Tess Simpson to the mathematician Heinz Hopf in Zurich who had become very concerned about Fröhlich's situation.

Fröhlich had also contacted the Emergency Committee in Aid of Displaced Foreign Scholars in the USA, and Hermann Weyl at Princeton.[12,272] However, no assistance could be provided from that route either. With no position in Czechoslovakia, Fröhlich was now on the poverty line. Reinhard Heydrich introduced a new law that all Jewish people in the German Reich and Protectorate should wear a yellow star from 1 September 1941. Fröhlich and his wife Elise were caught by the police on 19 September 1941 without the star and received a large fine, which they could not pay. Under Nazi rules, they now had a criminal record.[272]

On 21 October 1941, Fröhlich and his wife were deported to the Litzmannstadt Ghetto at Lodz in Poland in Transport B, just before Ludwig and Hedwig Berwald. As was the case with the Berwalds, the conditions were appalling for the Fröhlichs in Lodz.[272] Walter Fröhlich died in Lodz on 29 November 1942 at the age of 39. It seems his wife Elise also died in Lodz, although there are no records for her. As we have seen with several other academics who had applied to the AAC/SPSL, enquiries on Fröhlich's fate were made after the war, but it was not possible to find any details or trace any living relatives.

Both Berwald and Fröhlich applied to the SPSL, but it was not possible to save them. There were, however, a small number of Czech mathematicians who were successfully moved towards the end of 1939, helped by the SPSL.[267] Karl Löwner was a colleague of Fröhlich's (and also his referee) who did obtain a UK visa and an offer to go to Cambridge in 1939, but, like Fröhlich, the start of the war on 3 September 1939 prevented the move.[275] However, with the help of John von Neumann, he managed to obtain an appointment at Louisville University in the USA and left Czechoslovakia on 22 October 1939. He then took a berth on the liner *Statendam* from Rotterdam to New York. His mathematical career flourished in the USA during and after the war (he then used the name Charles), and he was eventually appointed to a professorship at Stanford.[276]

Another example is Arthur Erdélyi, who had been working in Brno, which is situated southeast of Prague. Born in 1908, he was younger than Fröhlich and also single, both features which made a move to the UK easier. He had corresponded with distinguished professors including Edgar Poole at Oxford and Edmund Whittaker in Edinburgh. Whittaker made available a small research grant which enabled the SPSL to make arrangements for Erdélyi to move to Edinburgh in February 1939.[277] There, he continued his very productive mathematical career.

5.3 Robert Remak

While working in Berlin, Robert Remak (1888–1942) made broad and significant contributions to mathematics and economics theory that have remained important to the present day. After a brief visit to the UK in 1939, he spent the early part of the war in the Netherlands, before being transported to Auschwitz.

Remak was born on 14 February 1888 in Berlin. He came from a distinguished Jewish family. His father Ernst was a well-known Professor of Neurology at the University of Berlin and his grandfather, also called Robert, was the first to propose the idea that cells are made by the division of pre-existing cells. From a young age, Robert Remak was interested in mathematics. He studied at the universities of Berlin, Marburg, Göttingen and Freiburg. He was supervised by the distinguished group theoretician Georg Frobenius. Remak's doctoral thesis, submitted in 1911, was on the decomposition of finite groups into irreducible representations. He married Hertha Meyer from Hamburg on 20 May 1926.

Remak's habilitation examination, however, was delayed by World War I, during which he served in the army. Even at a young age, he

Fig. 5.3.   Robert Remak (~1933).

developed a reputation of being a "difficult colleague," which caused complications in his career despite his mathematical brilliance.[278] He had a reputation for dressing informally and rudely correcting the lectures of the great mathematical and scientific names of the day in Germany, including Hilbert, Courant and Planck. He was also interested in the mathematical aspects of economics and was thought to have views with a communist tendency, which did not go down well with some of the more conservative mathematicians in Germany. He was eventually given a post to teach pure mathematics at the University of Berlin by Richard von Mises. In the early 1930s, he also made original contributions to algebraic number theory (see Figure 5.3).

After Hitler came to power in 1933, Remak was dismissed from his teaching position at the University of Berlin in September 1933 under the new Civil Service Laws for non-Aryans. However, his wife Hertha was Aryan and he felt, at least initially, that this would give him some protection. Nevertheless, on 22 March 1934, he wrote to the AAC asking for assistance:

> I would like to apply for an apprenticeship at a foreign university …
> I am handicapped in my scientific work because the university facilities
> are no longer available to me. I have private assets, the interest on which
> is sufficient to support me and my wife in Germany with modest living
> standards. Since I cannot see what part of my income would still be
> available to me if I stayed abroad for a longer period of time, I am
> unfortunately forced to raise the question of a financial subsidy for
> discussion.[279(t)]

This was Remak's first communication to the AAC/SPSL which made exceptional efforts to assist him over the next five years, as did several other UK organisations and individuals. These efforts are described here in some detail. The highly regarded Cambridge mathematician Hardy received a request about Remak from the AAC, "for an opinion concerning his standing and any suggestions you have to make regarding the possibility of finding a place for him. You will notice that he appears to be financially provided for." However, Remak's complicated personality was already known in the UK. Hardy wrote a somewhat cautious letter back to the AAC on 7 April 1934:

> Dr. Remak is a good and well-known mathematician who has done several important pieces of work … but I cannot put his case at all in the first line among those you have to consider, especially in view of the facts (a) that his previous position (if he had any regular position which does not appear clearly from the documents) was not an important one. (I suspect negligible from the financial point of view in comparison with his private income: but this is conjecture), (b) that he has private means, (c) that he is (this is a matter of "common gossip" in mathematical circles) highly eccentric and "difficult".[279]

As we have seen in other cases, a remark of caution from one of the leaders of a field in the UK at once made it less likely that assistance could easily be provided. On 3 November 1934, a formal application from Remak was received by the AAC. His extensive list of referees contained some of the most distinguished mathematicians, including Schur, Landau, Furtwängler, Löwner, Hopf, Speiser and Polya. He added the following eccentric note:

> I don't suffer from any material problems, but I do suffer mentally. Since the university facilities are no longer available to me, I am handicapped in my scientific work and, above all, I miss regular contact with mathematicians, which greatly reduces my ability to work. I am already looking ahead to go on a study visit abroad at my own expense. That failed because I had problems with the foreign exchange.
>
> I recently received an invitation to come to Cambridge, England, where some of my students are already, and where I hope to find particularly beneficial intellectual stimulation through an exchange of ideas with the group theorist Prof. Ph. Hall. However, because of the currency difficulties, I see no possibility at the moment of realizing my stay in Cambridge, and I would be very grateful if you could help me in any way.[279(t)]

The Assistant Secretary of the AAC, C. M. Skepper, clearly noting the problem with Remak's "difficult" character as mentioned by Hardy, replied firmly to Remak on 7 November 1934: "Unfortunately the funds of the Council are at present exhausted and we see no possibility of making it possible for you to come to England to accept the invitation from Cambridge."[279] Remak, however, would not take no for an answer and wrote again to the AAC on 15 November 1934: "I received an extremely kind private letter from Mr. Hall. I would be very grateful if you would intercede for me in Cambridge to have an official invitation sent."[279(t)] Skepper then wrote to Hall in Cambridge, who replied that he could not himself provide an official invitation and this would need to come from the Cambridge Faculty Board. Skepper then replied somewhat curtly to Hall saying that it would be best if the University of Cambridge made any arrangements with Remak independently as he was "not in urgent need." Remak, however, seemingly had not understood the tone of the letters from the AAC and wrote again on 29 January 1935 asking if assistance could be provided with an application to Professor Mordell in Manchester.[279(t)]

The AAC did not take any further action and Remak stayed in Berlin where his mathematical reputation continued to grow despite the darkening political distractions.[278] In 1936, the famous group theoretician Issai Schur, who was born in Russia and was a Professor in Berlin, stated that Remak was "undoubtedly the leading capacity in the beautiful and important field of the geometry of numbers."[280] However, during the infamous Kristallnacht of 9–10 November 1938, Remak was arrested and sent to the nearby Sachsenhausen concentration camp. Several people, including his wife, then worked very hard to obtain his release, and a series of urgent letters arrived at the SPSL.[279] On 22 November 1938, Dr. Hermann Blaschko from the German Jewish Aid Commission based in Cambridge wrote to the SPSL: "Dr. Remak is in a concentration camp and the family desperately needs to obtain some possibility for him to emigrate. If you can see any way in which you can help we will be most grateful." Tess Simpson from the SPSL then replied: "Do you know of anyone who could guarantee Dr. and Mrs. Remak's maintenance over here? I am afraid that is the only way of getting a visa for them."[279]

On 25 November 1938, a pleading telegram from Hertha Remak arrived at the SPSL: "Please help my husband Robert Remak and us to emigrate. This is now urgently necessary. I have been alone for fourteen days and am very desperate."[279(t)] Tess Simpson at once replied, "We need to find a British person who will guarantee maintenance for your husband and yourself for an indefinite period in this country." Hall from Cambridge then wrote to the SPSL on 3 December: "Dr. Remak is a distinguished mathematician ... he is at present, I believe, in a concentration camp. I should add that Dr. Remak has the reputation of being a little difficult to get on with. But I am sure you will agree that this is not a sufficient reason for refraining from helping him to escape." Mordell from Manchester also wrote with concern to the SPSL on 13 December: "Some of his acquaintances might be prepared to raise some sort of a fund on his behalf." The German Jewish Aid Committee then wrote to the most eminent of mathematicians, Hardy, at Cambridge, asking if some suitable post could be provided for Dr. Remak "as he has been in a concentration camp since the last pogrom."[279] Hardy replied from Trinity College, Cambridge, on 23 December with a more positive view than previously:

> You wrote to me on the 19th about Dr. Robert Remak. He was on the books of the Gordon Square Society. They did nothing for him then (I think rightly) because (a) he had private means and (b) he is an eccentric person unlikely to obtain permanent employment in England or the USA.
>
> Dr. Hahn, of Gordonstoun School, Elgin, a cousin of R's, has written to me saying that, if temporary provision could be made for him, he, Dr. Hahn, could undertake to have him sent to Bolivia after an interval. Exactly what that means I do not know. If it means that there could be a firm guarantee that his case would be settled in a reasonable time, then it would be possible to raise a small sum privately (£50 I should say certainly, possibly more) to tide over the interval. This, however, would be contingent on the guarantee. I do not think for a moment that the people willing to subscribe would be prepared to give any sort of permanent guarantee themselves. I have written again to Dr. Hahn.
>
> Ps. Dr. Remak is a mathematician of considerable distinction.[279]

The letter refers to Kurt Hahn of Gordonstoun School in Scotland, which was to have a pupil after the war who eventually became

King Charles III in 2022. Then, on 29 December 1938, Mordell wrote to the SPSL to say that he had heard that Remak was being let out of the Sachsenhausen concentration camp to go to the Netherlands. The SPSL was also contacted by the Cambridge Refugee Committee and the Birmingham Refugee Committee with concern about Remak. This, together with very tenuous guarantees from Hardy and Kurt Hahn, was sufficient for the SPSL to put in an application to the Home Office for a visa for Remak and his wife to enter the UK.[279] The visa was confirmed by the Under Secretary of State on 4 February 1939, and a letter was sent to Remak's home address at Manteuffelstr 23, Berlin, stating he should apply to the British Passport Office in Berlin to obtain the visas. Remak then wrote on 16 February 1939 to the SPSL from his home address with no mention of his incarceration:

> Thank you very much for your letter of February 6th and copy of the Home Office letter of February 4th, advising me of visas to England for myself and my wife. It has since been confirmed by a letter from the British Passport Control office in Berlin. I will get the visas as soon as I have received my passport, which has already been applied for. In order to be able to assess my prospects for eventual advancement in England, it is of the greatest importance for me to find out whether any guarantees for my subsistence were given before the visa was issued in England, if not by whom, by which private individual, association or university authority. I would be very grateful if you could let me know about this soon.[279(t)]

Kurt Hahn also wrote to the SPSL on 28 February 1939 stating that he had recommended that Remak should go at once to the Netherlands "which I understand he can enter on a limited permit." Then, the SPSL received a letter on 4 March from Colin Hardie of Magdalen College, Oxford, implying that some financial help might be possible from his College to support Remak and another mathematician, Otto Blumenthal. Hardie also stated that he had been in contact with Hans Freudenthal who was a distinguished mathematician from the University of Amsterdam and was to host Remak in the Netherlands. Magdalen College had set up a significant fund to assist academic refugees, which was already being applied for other subjects.[281] David Cleghorn Thomson, General Secretary of the SPSL, replied on 6 March to say that Remak was "a brilliant mathematician but we gather a little bit difficult to get on with in some ways." He also stated, "in the case of Blumenthal we know that his personality is very attractive ... Incidentally, I notice he is sixty-three, which will make his re-establishment somewhat difficult." Redvers Opie from Magdalen College then wrote to the SPSL to say that "we shall probably help Remak in the event of his being able to get to the Netherlands. We propose to do this through Freudenthal direct."[279]

Remak went to the Netherlands in March 1939, but a perturbed Freudenthal was at once writing to Heinz Hopf expressing concern that Remak was complaining to his landlord about the quality of his accommodation, which might force his expulsion back to Germany. Then, on 10 June 1939, Remak suddenly appeared in London and contacted the SPSL. He also visited Cambridge as was described by a surprised Hardy to the SPSL on 13 June:

> Dr. R. Remak called on me today, much to my surprise. I thought he was in Holland. I did not find it easy to follow him, but I understood him to say that (a) he has got a subvention (£200+) from Magdalen College, Oxford and, (b) he has been living in Holland on English money (I think supplied by Dr. Hahn, in anticipation of the Oxford money), (c) that he was too exhausted, after his experience, to work and needed a holiday which he proposed to take (for economy's sake) in Holland (returning there for the summer), (d) that he proposed to come to Cambridge for the winter (perhaps arriving Sept. or Oct.).
>
> All this seems reasonable enough but I would like to know what the facts are directly from you (and what has happened to the project of settling him in Bolivia or Panama).[279]

Cleghorn Thomson then wrote to Opie at Magdalen College on 15 June:

> I wonder if it would be possible for you to send me a further sum from the amount you have earmarked for Remak? He has been in here to see us and told us that you have already very kindly sent about £60 to

him on which he has been existing in Holland. I have written to his cousin, Kurt Hahn, to find out his ideas on the future and I had a long talk today with Professor Hardy of Cambridge who is another of his advisers. It is most important that he should be advised to stay in the most economical centre meanwhile, as neither Hardy nor I feel that there is much chance for a job for him in either Holland or England and consider it most likely that he will have to go to South America as originally planned. Meanwhile he requires money moving about the country and we gave him a fiver the other day. I forget what the total amount of money we voted for him is but seem to recollect that it was in the neighbourhood of £200.[279]

Opie replied on 16 June: "The Committee expected this grant to keep Remak and his wife for a year at the rate of expenditure estimated by Dr. Freudenthal in Holland. I do not believe that the Committee would be willing to make any further grant to Remak." Hardy also wrote to the SPSL to say that he had another meeting with Remak who had firmly stated that "going to Bolivia or Panama was impossible on climatic grounds." Hardy also said that Remak was attempting to obtain a visa to the USA. Then, on 27 June 1939, Cleghorn Thomson wrote to Hardy in somewhat exasperated terms:

Things are growing rapidly more complicated in the matter of Remak ... he really is in an advanced stage of paranoia. I tried to persuade him to return quietly to Holland at the end of this week ... but this he will not do and at present demands first a certainty that we approve of the settling in a house in Cambridge this autumn with his old furniture specially brought over from Germany ... When I asked him to go to Holland and await a letter from me he refused to do this and began to accuse us of taking part in conspiracy which had been going on for twenty years. If he were tucked away in Manchester with his wife and developed a Mathematical Society there, among people who understood his eccentricity, all would be well.[279]

Hardy had expressed reservations about Remak coming to Cambridge, so Remak went to Manchester to talk to Mordell about the prospects of working there. Mordell, however, was also not enthusiastic about this possibility and told Remak that he was going on a sabbatical to the USA the following year. On 10 July 1939, Mordell wrote to the SPSL: "He is obviously a bad case of nerves after his experience in a concentration camp and it would be probably best for them if he spent the next year in Cambridge." On 9 August 1939, Remak informed the SPSL, from a guesthouse in the Netherlands, that it was still his intention to come to Cambridge in September or October.[279]

Then, war was declared on 3 September 1939. On 9 September, Remak, perhaps now sensing the clear lack of enthusiasm from both Cambridge and Manchester, informed the SPSL, from Reinier Claeszenstraat 52 Amsterdam, that the war had changed his plans and he would stay in the Netherlands. His wife Hertha had joined him in Amsterdam on 31 July 1939, but after the start of the war returned to Berlin on 15 September 1939. She then initiated divorce proceedings, presumably to save her own life.[282]

In early 1940, the highly sensitive Remak was placed in a convalescent home in Amsterdam. The funds from Magdalen College, Oxford, continued to support him while he remained in the Netherlands right up to the time that the Germans invaded that country on 10 May 1940.[279,281] Then, all correspondence from Remak with the SPSL stopped. Subsequent reports in the Dutch police files show that Remak was reported to them for causing a disturbance by threatening residents in his accommodation on 12 June 1940, soon after the German occupation of Amsterdam.[283] The police files stated: "Dr. Remak had already been told by the immigration service that if he continues this harassment he will be expelled from the country."[283(t)]

The next the SPSL heard about Remak was a letter of 23 September 1946 from the UK Search Bureau for German, Austrian and Stateless Persons from Central Europe, stating that Remak had been deported from Amsterdam to Annaberg, Germany, on 10 November 1942 and nothing had been heard about his fate since then.[279] It was subsequently learnt that he was arrested on 3 May 1942 and held at the Dutch Transit camp of Westerbork until 10 November.[283] Of the more than 100,000 people who were interned at Westerbork, only close to 5,000 survived the war. Transports from Westerbork to the camp at Annaberg in Upper Silesia were common. However, more recent evidence suggests that Remak was taken straight to Auschwitz on 10 November with 757 others. The date of his death was reported as 13 November 1942.[284]

It has to be said that the SPSL and several other organisations and individuals did the best they could to help this "difficult" character, but his eccentricity contributed to his fate. In the 21st century, Remak's deep and broad mathematical contributions continue to be highly influential. His almost forgotten papers on the linear mathematical economics of national debts and superimposed prices have since emerged and found to be very relevant.[278] Remak has also given his name to finite algebraic number fields, group theory decompositions, units and regulators.[282]

5.4 Otto Blumenthal

Otto Blumenthal (1876–1944) was the first research student of David Hilbert. He was appointed to the Chair of Mathematics at Aachen at the young age of 29 and was editor of *Mathematische Annalen*. At the start of World War II, he was trapped in the Netherlands like Robert Remak and eventually died in Theresienstadt.

Fig. 5.4.   Otto Blumenthal as a student in Göttingen (left) and a professor at Aachen (right).

Ludwig "Otto" Blumenthal was born in Frankfurt on 20 July 1876. He was the son of the physicist Ernst Blumenthal. He studied mathematics at Göttingen and was the first doctoral student of David Hilbert. His teachers there included Felix Klein and Arnold Sommerfeld. Blumenthal's broad research publications in pure and applied mathematics covered topics such as complex function theory and the fundamentals of aerodynamics.[285] In 1905, at only 29 years old, he became the Chair of Mathematics at the University of Aachen, joining his friend Sommerfeld, who was the Chair of Applied Mechanics (see Figure 5.4). In the next year, he was made Chief Editor of *Mathematische Annalen*, an important appointment he would hold until 1936.[286] *Mathematische Annalen* was one of the leading journals in mathematics, with previous editors including Hilbert and Klein.

In 1908, Blumenthal married Mali Ebstein. He served in World War I as head of a military weather station and was also involved in aircraft construction, for which his mathematics knowledge was useful. Shortly after Hitler came to power, Blumenthal was denounced by students at Aachen as a communist on 18 March 1933. He was briefly taken into custody by the Gestapo.[285] He was then dismissed from his chair on 22 September 1933 for membership of international pacifist organisations such as the German League for Human Rights.[286] He applied to the US Emergency Committee in February 1934, and his application to the AAC was received on 3 September 1934, in which he stated that his research interests were in Analysis (Theory of Functions) and Hydrodynamics.[287,288] In his AAC application, he wrote that he had Jewish parents and was a member of the Protestant Church. His referees included Hilbert, Sommerfeld, Hardy, von Karman and Harald Bohr. As with other applications from mathematicians, Hardy at Cambridge was asked for his frank opinion on Blumenthal. In his reply on 13 October 1934, he questioned whether Blumenthal wanted to leave Germany as he was still editing the *Mathematische Annalen*. Hardy said he was one of the major characters in German mathematics but rated him as a "B".[288] As we have seen before, a lukewarm assessment from a leading UK academic would quickly dull the possibility of support from the AAC/SPSL.

In November 1938, Blumenthal was fired from his editorship of the *Mathematische Annalen* following a decree instigated by the Nazi Minister

Bernhard Rust that removed Jews from membership of scientific societies. However, Springer, the publisher of the journal, continued to pay a pension in recognition of Blumenthal's editorship of 31 years and he managed to divert these payments to his sister.[285] In the meantime, Blumenthal's son Ernst and daughter Margrete had both moved to study in England.

No more action was taken until 5 February 1939 when a letter was sent to the SPSL from the influential J. H. C. Whitehead at Oxford supporting Blumenthal.[288] Blumenthal had corresponded with Whitehead on his deep interest in the history of Greek mathematics and Whitehead had then written to the Bishop of Chichester to see if some support could be obtained for him in the UK from this novel angle. Paul Ewald, a friend of Blumenthal who had recently taken up a lectureship in mathematics at Queen's University Belfast, also wrote to the Bishop. Shortly after this, on 4 March 1939, Colin Hardie had sent a letter to the SPSL from Magdalen College, Oxford, stating that it may be possible to provide some financial support for Remak and Blumenthal.[279] Just as in the case of Remak, David Cleghorn Thomson, the SPSL General Secretary, replied at once to Hardie stating that Blumenthal's age would make a placing difficult. Once again, this emphasised the challenges in finding a suitable position for the most senior scholars even if they had an outstanding track record of publications. Accordingly, Redvers Opie in his reply from Magdalen College on 9 March stated that it may be possible to find financial support for Remak, but did not mention Blumenthal.[279] The Bishop of Chichester also wrote to the SPSL about Blumenthal, and Tess Simpson responded on 14 April 1939: "He is very highly recommended, but the difficulty is his age – sixty-three ... If the Bishop himself were interested, he might be able to think of some way of helping Professor Blumenthal."[288] That was the last letter from the SPSL on Blumenthal until after the war.

In the meantime, Blumenthal had moved with his wife Mali to Utrecht in the Netherlands on 13 July 1939 with the assistance of Freudenthal.[285] He then managed to obtain a short visa to visit his son and daughter in England in August. However, with the strong possibility emerging of a war between Germany and the UK, his daughter Margrete encouraged him to return quickly to the Netherlands where she assumed that the situation would be safer.[285] This turned out to be a fatal mistake. With the capitulation of the Netherlands to the Nazis on 15 May 1940, the situation there became very difficult for Jewish people — especially those who had left Germany. His German citizenship was formally annulled on 11 January 1941, making him stateless. After being forced to move several times, Blumenthal and his wife Mali were placed in the Westerbork internment camp on 22 April 1943, after which Mali died just one month later, on 21 May.[284,285] After hearing that his sister, Anna Storm, had been transferred to Theresienstadt, Blumenthal, who was now alone, asked for a transfer there and he was transported on 18 January 1944.[285] On arrival, he discovered that his sister had died some months earlier. As we know, Theresienstadt was a German-speaking camp with several intellectuals already incarcerated there. Despite the atrocious living conditions, Blumenthal somehow managed to give some lectures in mathematics to an interested audience. His health, however, was failing and he died on 12 November 1944 of natural causes that included pneumonia and tuberculosis.[284,285]

One should also mention Hans Freudenthal, the influential mathematician who had brought Blumenthal and Remak to the Netherlands. Being Jewish, he was dismissed from his lectureship at the University of Amsterdam after the Nazi invasion. He went into hiding with his family and survived the war. He then took up a professorship at the University of Utrecht, promoted mathematics education and wrote widely on the history of mathematics. He died on 13 October 1990 at the age of 85.[289]

In the same way as Fröhlich, Berwald and Remak, the mathematical discoveries of Otto Blumenthal continue to be important to the present day. He gave his name to many mathematical results. His Blumenthal Theorem on densely distributed zeros of orthogonal polynomials has been widely applied. His work with Hilbert on modular forms has also been influential. His more applied research in extending the Stokes–Helmholtz theorems with vector analysis has applications in a variety of fields, including fluid mechanics and electrodynamics.[285,286]

Blumenthal is remembered in Aachen where he held his professorship. A street is named after him, together with a commemorative plaque, and there have been several articles written in Aachen about his life and career.[290] Max Born's poignant list of 35 family members or colleagues who were victims of the Nazis included Otto Blumenthal.[124]

Not all distinguished mathematicians who were unsuccessful in their applications to the AAC/SPSL and sent to Theresienstadt perished during the war. Paul Funk (1886–1969) was an Austrian mathematician who worked on the calculus of variations and introduced the Funk Transform. Like Blumenthal, he was a student of Hilbert. He became a professor at the German University in Prague where his colleagues included Fröhlich and Berwald. His application to the SPSL was received very late, on 31 March 1939, and he had exceptional referees, including Prandtl, Weyl, von Karman, Mises and Levi-Civita.[291] On 28 April 1939, Hermann Weyl wrote from Princeton:

> Funk and I were students together in Göttingen. It would never have occurred to me that he was non-Aryan. In appearance, manners and speech he struck me as a true son of the alps, rough and cheerful ...
>
> Much of his work is in applied mathematics (elasticity, electric currents, optics). He is particularly fond of teaching and ever so ready to put his mathematics at the disposal of any mathematician or engineer who comes around with a nice problem. In this way he has made himself a most useful member of the Technische Hochschule in Prague. But it is partly due to such distractions that his mathematical output does not show the full stature of his talent.[291]

This typically honest reference from Weyl, although given with the best intentions, meant that the SPSL executive committee would not give Funk priority in the very competitive atmosphere of 1939. Funk stayed in Prague and was transported to Theresienstadt towards the end of 1944. He was one of a small number of survivors and lived for another 25 years, becoming a professor at the Technical University of Vienna after the war.

Leading mathematicians were hit particularly hard by the Nazis.[272,280,292] Many were pure mathematicians who could only be assimilated in a very small number of specialised research groups in leading universities overseas. Aside from those (Berwald, Fröhlich, Remak and Blumenthal) who sought, but ultimately did not receive, assistance from the AAC/ SPSL, there were several other distinguished central European mathematicians who perished through the actions of the Nazis. These included George Pick, a colleague of Berwald and Fröhlich in Prague, Felix Hausdorff from Bonn, Alfred Tauber and Ludwig Eckhart from Vienna, Paul Epstein from Frankfurt, Fritz Hartogs from Munich, Paul Lonnerstädter from Würzburg, Nelly Neumann from Essen, and Kurt Grelling, Charlotte Hurwitz, Margarete Kahn and Reinhold Strassmann from Berlin.[280] There appears to be no evidence, however, that these mathematicians applied to UK organisations for assistance.

As has been emphasised in this chapter, there are some similarities that have emerged from our analysis of the prospective refugees in chemistry and physics and the mathematicians who interacted with the AAC/SPSL. Age and seniority were negative factors, with younger refugees being easier to relocate. Some of the mathematicians seemed to be almost too eminent to be placed. A slight reserve or negative remark from the experts in the UK made a placement unlikely and the mathematicians had to have the right fit in a prospective university department. Pure mathematicians did not have the prospect of an industrial appointment, which was sometimes a possibility for a chemist, although the mathematicians did occasionally have a teaching option, which was not very popular. In addition, pure mathematicians often had eccentricities that appeared to let them down at crucial times, as seen in the case of the "difficult" Remak, and the obliviousness of Berwald in not realising the seriousness of the unfolding events. There was also just bad luck in timing, such as when visas were finally issued to both Fröhlich and Remak in the summer of 1939 but had to be cancelled following the start of the war. It seems that staying in the Netherlands during the war, as was the case for both Remak and Blumenthal, eventually became more dangerous than remaining in Belgium, where the chemical physicists Lasareff, Goldfinger and Rosen managed to survive.

During the war, the mortality rate of Jewish people was 40% in Belgium and 75% in the Netherlands.[293] There are several reasons that have been put forward to explain this difference. One factor is that the Netherlands was neutral in World War I, while Belgium built up a significant resistance movement which was reactivated in World War II, and indeed was joined by Lasareff, Goldfinger and Rosen, whose scientific knowledge in physical chemistry was put to good use. Another feature is that the civil service, the police force and the Jewish councils were more closely run by the Nazis in the Netherlands than in Belgium.[293-295]

CHAPTER SIX

Refugees in Medicine

After Hitler came to power in 1933, the British Medical Association made clear that it did not support a significant influx of medical doctors from Germany as this would diminish the employment opportunities for British doctors.[296,297] This view had a strong influence on the Home and Labour Offices which gave the approvals for visas and work permits to come to the UK. Another factor was the need to obtain the appropriate qualifications to practise medicine in the UK even if the person had previously had a distinguished medical career in their own country. These aspects made it difficult for refugee medical scientists to come to the UK in the 1930s.

However, a significant number of biomedical scientists who had published their research in learned journals were, like the chemists, physicists and mathematicians, assisted by the AAC/SPSL to come to work in the research institutions and universities in the UK.[297] Included among this group were Hans Krebs, Ernst Chain and Bernard Katz who would all go on to win the Nobel Prize for Physiology or Medicine. However, there were still some highly eminent biomedical researchers who applied to the AAC/SPSL but were not able to come to the UK. A large number of applications came into the AAC/SPSL which were classified in the many sub-areas related to medicine, including anatomy, physiology, pathology, pharmacology, paediatrics, radiology, cancer, cardiology, endocrinology, haematology, neurology, ophthalmology, psychology, bacteriology, orthopaedics, dentistry, otology, dermatology, forensic medicine, gynaecology and surgery.[298] In the following, we describe some of the most prominent examples of medical academics about whom the SPSL made enquiries after the war and who had tragic fates.

6.1 Arthur Simons

Arthur Simons (1877–1942) gave his name to Barraquer–Simons syndrome, which links the loss of fat tissue to tuberculosis. Working in Berlin, he also pioneered the use of film in medical studies. He was murdered by Nazi collaborators near the Baltic Sea.

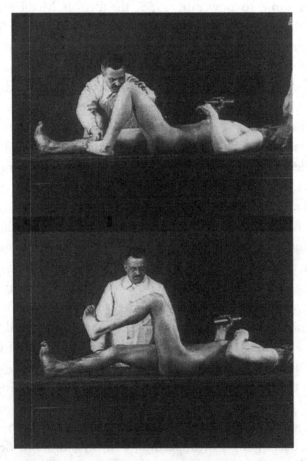

Fig. 6.1   Arthur Simons, stills from his film *Head Positions and Muscle Tone* (1923).[300]

Simons was born on 11 October 1877 to a Jewish family in Düsseldorf. He was the elder brother of the biologist Hellmuth Simons, whose extraordinary story is discussed in a later chapter. He studied medicine in Leipzig and Berlin. From 1907–1914, he was assistant to Professor Hermann Oppenheim in Neurology in Berlin. He became an expert on lipodystrophy, a condition associated with loss of fat tissue in certain parts of the body. He described a disease which is now called Barraquer–Simons syndrome which related lipodystrophy to tuberculosis.[299] During World War I, he was a medical officer and published on his observations of soldiers with serious neck and brain injuries. He recorded some of these observations with short movies, which was quite original at that time (see Figure 6.1).[300,301] After the war, he became a Professor Extraordinary at the Charité Hospital which was associated with the University of Berlin, where he also had a private medical office. He lived in a fine house at 50 Kurfürstenstrasse, Berlin, which was almost next to the house owned by Hedwig, the wife of Alfred Byk.

Following Hitler's rise to power, Simons' privilege to teach was withdrawn on 6 September 1933. On 15 November 1934, he applied to the Academic Assistance Council at the somewhat senior age of 57.[302] His referees included two neurology professors based in New York: Bernard Sachs, brother of Samuel Sachs who was a joint founder of the investment bank Goldman Sachs, and a former colleague in Berlin, Leopold Lichtwitz. He listed over 20 publications, several being on lipodystrophy. Following the usual procedure, Simons' application was forwarded by Cleghorn Thomson of the AAC to one of the leaders in his field in the UK, in this case Professor Sir Hugh Cairns of the Radcliffe Infirmary, Oxford. Cairns had established a major reputation in the treatment of head injuries and was the surgeon who would be called to treat T. E. Lawrence (of Arabia) after his fatal motorcycle crash in 1935. He is credited with the introduction of crash helmets for motorcyclists in Britain, thereby saving many lives. Cairns was also a key person who persuaded Lord Nuffield to make a major donation that transformed medicine in Oxford.[303] However, Cairns wrote on Arthur Simons: "I don't think he is an outstanding neurologist."[302] As we have seen with other scientific refugees, such a lukewarm comment was not going to give

Simons priority for assistance from the AAC/SPSL, and no action was taken.

In the meantime, Simons was being denied many of his basic rights in Berlin. He managed to retain some private patients, but in 1938 his licence to practise medicine was removed. He was allowed to be a "Krankenbehandler" to give medical care to Jewish patients. To make ends meet, he had to sell much of his private property.[299,300]

On 21 September 1943, Tess Simpson wrote to Bernard Sachs in New York, who had been one of the listed referees of Simons, asking if he had any knowledge on his whereabouts and saying that she sincerely hoped that he "managed to escape from the continent."[302] There is no record of a response. After the war, Ilse Ursell wrote to the UK Search Bureau for German, Austrian and Stateless Persons for Central Europe. A reply was received on 26 November 1946 that was typical of the post-war uncertainties:

> According to his former housekeeper Maria Schiffer of 50 Kurfuerstenstrasse, Berlin he did not return when he was ordered to see the Gestapo in September 1942. In April 1943, on the other hand Mrs. Schiffer received word from Dr. Simons' former assistant, Miss Elise Klein, that he was shortly to be deported to Reval. Later he was reported to be working in an epidemic hospital. There is no further trace of either Dr. Simons or Miss Klein.[302]

It was learned subsequently that Simons was arrested by the Gestapo in September 1942.[299] In the summer of 1942, Auschwitz-Birkenau had reached the brink of its killing capacity with the transport and murder of Western European Jews from the Netherlands, Belgium and France. Transportations from Berlin to the Minsk area, as was used for the colleagues of Marie Wreschner, were also filled to capacity as were transports to the camps at Theresienstadt, Treblinka, Belzec and Sobibór. Heinrich Himmler had proposed using Jewish people for the extraction of oil shale in the Baltic area and Adolf Eichmann then made the decision at short notice to introduce transports to that region.[120,304] On 26 September 1942, Simons was deported from Berlin to Raasiku in Estonia. The Reichsbahn train from the Moabit freight yard had 895 women and

354 men, with 108 children under the age of eleven. The dreadful journey in a cattle wagon took four days and nights.[304] On arrival at the Raasiku railway station, a selection process was carried out with about 200 younger people sent to labour camps. The remaining people, including the 65-year-old Arthur Simons, were taken to a dune and forest area near Kalevi-Liiva on the Baltic Sea where they were shot by Estonian collaborators under German supervision.[304] The murdered bodies were then covered in sand. Two years later in 1944, as the Soviet front approached, the pits were opened by the Sonderkommando and the corpses were burnt to cover up the mass killings. These graves at Kalevi-Liiva were only rediscovered in 1961.[305]

The work of Arthur Simons on neck reflexes has become increasingly important for "acute neurology and in rehabilitation therapy of hemiplegics for antispastic positions."[300] His pioneering films used for medical investigations were only re-discovered in 2010, nearly 100 years after they were made.[301]

6.2 Otto Sittig

Otto Sittig (1886–1944) published several papers on diseases of the nervous system in the German University in Prague. He was one of the last people to be murdered in Auschwitz.

Sittig was born in Prague on 7 September 1886. After studying in that city and graduating as a doctor, he became an assistant physician at the German University in Prague working with Alfred Pribram and Arnold Pick on aphasia. He served in the Austro-Hungarian army as a physician in World War I. After the war, he worked briefly in Berlin and Frankfurt before returning to the German University in Prague where he was eventually given the title Professor Extraordinarius of Neurology and Psychiatry. He published over twenty papers on diseases of the nervous system. He also had a private appointment at a sickness insurance company dealing with neurological cases, a position he retained until forbidden to do so by the authorities in 1938. In 1935, he had the distinction of being elected as a corresponding member of the Société de Neurologie de Paris.

Fig. 6.2   Otto Sittig in his SPSL application of 1938.

Following Hitler's entry into Prague on 16 March 1939, Sittig wrote to the SPSL enquiring about the possibility of moving to the UK (see Figure 6.2).[306] As we have seen with mathematicians from Prague, this was a very late application, making it difficult for assistance to be provided. His referees included Ludwig Guttmann of Balliol College, Oxford. Guttmann, who would become well known for establishing the National Spinal Injuries Centre at Stoke-Mandeville Hospital, had only just arrived in the UK himself after escaping from Breslau via the unusual route of Portugal where he had been requested to treat a close friend of the dictator Salazar.[307,308] Guttmann indicated his strong support for Sittig to the SPSL. Another referee, Sir Henry Head FRS, had made new discoveries on the mechanism of the nervous system by conducting experiments on himself in which he severed and reconnected nerves and studied how they regained their sensation.[309]

A letter was also written on 16 May 1939 by E. Arnold Carmichael who was a registrar on neurology working at the National Hospital for Nervous Diseases in Queen Square, London. He stated, "Of course, I will be only too glad to help Sittig in any way; but here again one is in the awkward position of having to say he is past his peak of effectual contributions to science."[306] Also, at the age of 53, Sittig did not have the advantages of youth and this, combined with the comments from Carmichael, would have normally meant that little would be done by the SPSL to help Sittig. He continued to publish up to the year of 1939, including an article in the first volume of the *Lancet* on a case of polio.[310]

On 25 April 1939, the SPSL received a letter from Mrs. Trude Sittig in London about her brother-in-law Otto. She stated she would be able to provide a sum of £200 as a guarantee during his stay in England. In addition, she included a letter stating that Otto Sittig would be provided with a visa to the USA by the American Consul in Prague during the next year.[306] This was sufficient support for Tess Simpson to write at once to the Home Office requesting a visa for Otto Sittig and his wife Irma (née Müller), stating that he would not be requesting employment in the UK. This request was granted and Tess Simpson was able to send a letter on 1 June 1939 to Sittig with the good news.[306]

He responded with thanks on June 8 and, perhaps not realising the restriction on employment, stated that he would "like to continue my researches on apraxia, on the grasp reflexes of infants and on pupillotony. Another subject that would interest me would be the early diagnosis and treatment of poliomyelitis." He also sent another letter the next day stating, "I am taking all the necessary steps for emigration, but there are many formalities to fulfil, which take presumptively a considerable time."[306] As was the case with the mathematician Walter Fröhlich, also from Prague, there was an unfortunate delay in obtaining the visa for Sittig's wife Irma and Tess Simpson wrote to the Home Office on 2 August to attempt to accelerate this process. However, the visa did not come through before the start of the war on 3 September 1939 following which all visas were cancelled. Then, on 7 October 1940, before the USA entered the war, Trude Sittig wrote to the SPSL requesting for the return of the documents that the SPSL retained on her brother-in-law stating, "He was prevented from coming to this country due to the outbreak of war and I would like to get in touch with the American sponsor."[306]

In October 1943, Tess Simpson wrote to several of Sittig's referees asking if they had an address or other information about him. Ludwig Guttmann replied on 20 October: "As far as I know, he did not succeed to get away from the Continent, and was able to continue his work in Prague during the first year of the war."[306] That was the last communication the SPSL had on Sittig.

On 8 September 1941, Sittig had been forced, with over 60 other Jewish people from the Prague region, to hand over a considerable "compensatory tax" of 3537 marks.[311] Then, on 5 July 1943, he and his wife Irma were transported to the notorious Theresienstadt ghetto in Czechoslovakia. Finally, on the late date of 28 October 1944, this eminent neurologist and Irma Sittig were transported to Auschwitz where they were murdered at Birkenau.[67] He was 59 years old. There were over 2,000 people on this transport, which was the final one from Theresienstadt to Auschwitz, and use of the gas chambers in Birkenau was stopped just five days later.[22,120] A total of 169 people from the transport survived the war. After this last transport to Auschwitz, there were just over 11,000 people left in Theresienstadt.[69]

6.3 Erich Aschenheim

Erich Aschenheim (1882–1941) made several important contributions to public health including setting up a pioneering baby clinic in Remscheid, Germany. He took his own life in 1941.

Erich Aschenheim was born in Berlin on 4 February 1882 to a family of Jewish descent. His father Leopold was the director of an electricity supply company in Berlin. His uncle Sir Felix Semon was elected President of the British Laryngological Society and was physician to King Edward VII. Aschenheim studied medicine in Munich and Berlin and received the licence to practise in 1907.[312,313] In the same year, he received his doctorate for research on multiple aortic aneurysms. He started to publish papers on haematology and took up positions first at the University Hospital in Heidelberg in the infant department, then in Dresden and, shortly before the start of World War I, in Düsseldorf where he was an assistant to Arthur Schlossmann who had founded the world's first special clinic for babies. In 1906, Erich Aschenheim married the art historian Emma Ehrmann and they had three daughters. She died in 1910.

During World War I, Aschenheim served as a medical doctor. In 1919, he married Annemarie Appelius, with whom he had a daughter. Following the war, he gave lectures on skin, nervous and mental diseases in infants. He also set up a new baby clinic, a home for small children with rickets and several other public health initiatives in the city of Remscheid. He continued to research and published over 60 papers in areas related to public health. He also published a major book on this topic in 1927.[313]

Following the rise to power of the Nazis, Aschenheim was fired from his posts in Remscheid on 22 March 1933, even though he had served as a medical officer in World War I. Remscheid had a strong middle-class base of ardent Nazi supporters which made things difficult for a Jewish doctor. Aschenheim then moved to Krailling in Bavaria near Munich where he continued to practise paediatric medicine. However, his licence to practise was eventually revoked on 30 September 1938.[312]

Aschenheim's application to the AAC was received on 10 November 1934.[314] In his application, he wrote, "I ask you not to send any inquiries to my referees unless there is a secure prospect of employment." As references from leaders in a field were vital for applicants to the AAC to gain support in the UK, this put him at a significant disadvantage despite his impressive publication record. Furthermore, he could not obtain a reference from Professor Schlossmann who had died in 1932. However, his uncle Sir Felix Semon did forward a strong summary of Aschenheim's achievements to the AAC. Aschenheim also wrote that he had taken up the Protestant faith.[314]

Aschenheim's application was sent to several experts in child health in the UK. Sir Francis Fraser of St. Bartholomew's Hospital had written on 21 December 1933: "I am of the opinion that Dr. Aschenheim's chief interests and abilities lie in the direction of Social Hygiene rather than in Clinical Paediatrics, Clinical Medicine or laboratory work, and that he is not now suitable for research work on diseases and their causation."[314] As we have seen from applications in other subjects, an opinion which was negative on research prospects was going to make it very difficult for Aschenheim to find a placement in a UK university or research institution. This view was confirmed by Charles McNeil from Edinburgh who also said that he thought Aschenheim did not have the appropriate qualifications required to practise as a doctor in the UK despite his impressive clinic record. Aschenheim's papers were then sent to Alexander Carr-Saunders at the School of Social Science at the University of Liverpool, who, four years later, was to be made director of the London School of Economics and knighted. Carr-Saunders responded that there were no courses in social hygiene in universities in the UK which Aschenheim could teach. Accordingly, Aschenheim did not tick any of the important boxes and the AAC was unable to facilitate his application, despite the strong support from his uncle Sir Felix Semon.[314]

In November 1938, Aschenheim was arrested by the Gestapo in Düsseldorf. His father-in-law managed to arrange for his release after twelve days in custody but he was placed under police surveillance.[312] With his application to the UK being unsuccessful, he applied through a Protestant pastor to become a missionary doctor in South America but there was no luck there either. On 4 May 1941, Erich Aschenheim took his own life in Krailling with cyanide.[312,313] It was not his first attempt at suicide.

In December 1945, the SPSL wrote to the Central Tracing Bureau which was run by the US Army in Wiesbaden, asking for information on Aschenheim.[314] The Bureau replied in February 1946 that he had committed suicide but his wife Mrs. Annemarie Aschenheim, who was not Jewish, and one daughter were still living at their old address in Krailling. Mrs. Aschenheim had written on 13 January 1946 to the Tracing Bureau:

> My husband is dead. He was a prisoner of the Nazis in 1938 after the Putsch of the Jews and synagogues. He lost his medical appropriation and he was even tortured by the Gestapo and the Nazis. He had tried for a visa in England in 1939 because the daughters of him and his brother were in England since 1938. But the visa would not come from

England, though the Bishop of Chichester was very interested in my husband. In 1941, in a short time before he was to come to a K. L. (Concentration Camp), he killed himself because he could no longer suffer the torture of the Nazis.

I lived with my daughter for the last half year and we are both at the end of our health and our soul strength ... Is it possible that I can write through the UNNRA to my daughters who are living in London?[315]

The message was passed on to the UK Search Bureau for German, Austrian and Central European and Stateless Persons to see if it had any information on the two daughters, Anni and Gabriele, who were living in Britain. The Search Bureau was able to trace them and informed Mrs. Aschenheim and the SPSL that Anni was living in a Protestant Community in London and Gabriele was training to be a District Nurse in Gateshead.[315] The Bureau also said, "Both are very happy in their work. They hope to write as soon as the postal service is open again." On 1 October 1956, Annemarie Aschenheim initiated a restitution claim in Munich as a compensation for the treatment her husband had received.[316] She wrote with bitterness:

My husband — like all Jewish doctors — was deprived of his medical licence in October 38 according to the law. This banned the practice from continuing. A transfer to a non-Jewish colleague was no longer possible at that time. My husband recommended his circle of patients, who had already dwindled by then, to go to Dr. Wolfgang Stoeger in Planegg. Dr. Stoeger was the only one of his colleagues who had always behaved in an exemplary manner towards my husband, and later, when my husband attempted suicide and committed suicide three months later, he also took care of us in a humanely way.

This prompted me, after my husband's death, to give him (Dr. Stoeger) the part of his instruments that he could use. My husband's actual successor — without buying the practice — was Dr. Engelberg, who settled down as a doctor in Krailling and very soon got all the health insurance funds, since no other doctor worked in Krailling. He took over most of my husband's previous patients. However, neither during my husband's lifetime nor later did we have a claim against Dr. Engelberg to replace the practice. Under the circumstances at that time, we might have been able to raise a claim, but it would not have gained legal force.

In addition, thanks to the nasty agitation of a fanatical National Socialist colleague in Planegg, my husband's practice, which was initially extremely successful, had greatly diminished, so that Dr. Engelberg hardly seemed to us to be subject to recourse. The population was frightened by the machinations of this man. He then tried to buy my silence in the tribunal, which I refused. The real damage done to my husband's practice is due to this gentleman.[316(t)]

Eva, the daughter of Erich and Annmarie Aschenheim, who had stayed in Germany, followed in her father's footsteps. She trained as a psychologist in Munich and wrote scientific books and papers on juvenile delinquency. She died in Munich in 1975 aged 52.[313]

6.4 Ferdinand Blumenthal

Ferdinand Blumenthal (1870–1941) was an internationally regarded expert in cancer research who set up several leading medical institutes in Berlin. He was killed in a Luftwaffe air raid in Estonia in 1941.

Blumenthal was born in Berlin on 5 June 1870 to a Jewish family. He was not closely related to the mathematician Otto Blumenthal. Like the other medical scientists discussed in this book, he studied at several centres including Freiburg, Strasbourg, Zurich and Berlin. His doctorate in Freiburg was on the fundamental topic of the influence of alkalis on the metabolism of microbes.[317] Like Arthur Simons, he then worked at the Charité University Hospital in Berlin on bacteriology and biochemistry. He published several papers on diseases including cholera, diphtheria and tetanus. In 1905, he was appointed to an associate professorship at the Friedrich-Wilhelms University in Berlin.[318] There, his research turned to the study of cancer and in 1910 he published a major book on chemical processes in cancer. He became director of the Institute for Cancer Research and was not drafted into the forces during World War I due to the importance of his work. He became the leading cancer specialist in

Fig. 6.3   Ferdinand Blumenthal in 1930.

Germany and in 1919 was made secretary-general of the German Central Committee for Cancer Research and was editor of the German-based *Journal of Cancer Research*. He was a pioneer on developing techniques for the early detection of cancer and also in follow-up treatment of cancer patients.[318]

In his institute in Berlin, Blumenthal was instrumental in setting up other departments in X-ray analysis and a chemical laboratory, and he was at the height of his influence (see Figure 6.3). He also formed a department of haematology with Hans Hirschfeld, who is discussed later in this chapter, and a department of virology with Ernst Fränkel, who subsequently managed to escape to England. As was the case with all the scholars discussed in this book, everything changed when Hitler came to power in 1933 and Blumenthal was dismissed from his post. His outstanding institute with its impressive sub-departments fell apart. His dismissal even caught the notice of a publication in New York which commented on his dismissal together with that of others from Berlin, including the chemists Haber, Freundlich and Polanyi.[319] With a headline "More Wholesale Dismissals of World-Famous Scholars ordered by Nazi Minister," the *Jewish Daily Bulletin* of 3 May 1933 reported:

> One more German Jewish Nobel prize winner, twenty-one professors at the University of Berlin, nine at the University of Cologne, a world-famous surgeon and a world-famous neurologist, were dismissed from their posts today as Bernhard Rust, Nazi Minister of Education for Prussia, continued the work of eliminating the 'Jewish influence' from German cultural centers.
>
> Meanwhile systematic instructions were given to libraries for preparing books of Jewish authorship for the auto-da-fé on May 10. All libraries have been instructed to turn over such books to the authorities, and special squads of Nazis are touring the residential districts, collecting "fodder" for the flames. Students have been instructed to turn in all text-books that are of Jewish authorship. Universities also extended their boycott on Jewish culture into the fields of commerce, forbidding the purchase of ink, paper, and other school supplies from Jewish shops, for fear that the taint of Jewish culture might reach the students through such contact ...
>
> Prof. Fritz Haber, Head of Physical Chemistry in the Kaiser-Wilhelm-Institut of Berlin, and winner in 1918 of the Nobel Prize in Chemistry, headed today's contingent in the exodus of Jewish intellect from German institutions of learning. He is famous for invention of a process of obtaining nitrogen from the atmosphere. Two assistants, Prof. Freundlich and Polanyi, resigned with him ... Professor Ferdinand Blumenthal specialist in cancer research, and Prof. Friedrich Friedmann, specialist in tuberculosis, and Prof. Victor Jollas of the department of biology of the Kaiser-Wilhelm-Institut were among the dismissed scholars.[319]

Blumenthal moved briefly to Switzerland and then on to a position in Yugoslavia as a visiting professor in Belgrade's medical facility.[320] There, the arrival of Germany's leading cancer specialist received much publicity and he was invited to set up a new institute for cancer research modelled on his achievements in Germany. Almost at once, he was representing

Yugoslavia at the first International Cancer Conference in Madrid (now known as the Union for International Cancer Control).[320]

Blumenthal continued to publish on cancer topics but the atmosphere in Belgrade became negative for him. In Belgrade, there was much anti-Semitism and he also was a recipient of envy as he was a distinguished foreign medical scientist coming to a country with much acclaim. The newspaper *Narodna Odbrana* ran a campaign against Blumenthal, criticising the German language used in his lectures and examinations and stating, "Our university has welcomed with open arms every Jew the Führer of the Reich had banished from a strong, powerful and cultured Germany."[320] However, he did manage to arrange for the re-publication of his book *Results of Experimental Cancer Research and Cancer Therapy* in Belgrade.

Blumenthal's unhappiness in Belgrade had first been mentioned to the AAC in 1934.[321] Then, at the end of 1936, he moved to Vienna. This was a naive choice with the inevitable political storm that was coming to Austria. Following the Anschluss in March 1938, the prominent Blumenthal was arrested by the Gestapo and he spent three months in prison. This was almost at once communicated to the SPSL in urgent terms. On 21 April 1938, Anita Warburg, who was a prominent member of the Jewish Refugee Committee and British Red Cross, telephoned urgently to say: "The great cancer expert was now in a concentration camp. Can we get him an invitation over here so as to get him out? He has money outside Austria, quite enough to live on indefinitely."[321] Wolfgang Foges, the book publisher who had emigrated from Vienna, then followed this up with a letter to the SPSL on the same day:

> On the Friday night when the German troops marched into Austria, Professor Blumenthal was arrested by the S.A., though he is more than 70 years of age. Up until now it could not be ascertained why he was arrested, nor could anybody communicate with him. As the professor has never worked politically his family has only one possible explanation for his arrest: that he was called to the previous Chancellor Schuschnigg once as a doctor and that such a record was possibly found. All steps taken by the Hungarian and Yugoslavian Embassies in Vienna were unsuccessful. Therefore, the only way to save the professor, who

has ample means at his disposal, would be according to his friends to summon him to England for lecturing or teaching.[321]

Walter Adams, the General Secretary of the SPSL, then at once sent on the letter from Wolfgang Foges to Dr. William Cramer of the Imperial Cancer Research Fund and asked if Professor Blumenthal was "important enough" to justify sending an immediate invitation to come to the UK and also requested suggestions for institutions which may be able to send such an invitation. Cramer immediately telephoned the SPSL to say that he was asking the Royal Society of Medicine to invite Blumenthal to lecture, but "they cannot of course offer Blumenthal a job." Blumenthal's daughter, Herma, then wrote on 4 May 1938 from Prague to the SPSL to say that she could arrange for a financial assurance for her father if he was to move to the UK. Walter Adams at once responded to say that more concrete details of the financial assistance would be needed and under no circumstances would her father be able to work in the UK.[321]

The SPSL then received a letter from Blumenthal himself written on 18 June 1938 from the Hotel Palas in Belgrade:

> I have heard from Professor Ernst Fränkel in London that Dr. Gye and Dr. Adams have been trying to have me released from prison in Vienna. These efforts have been, as you see from this letter, successful. As I do not know the exact address of both these gentlemen I would ask you to convey my gratitude to these gentlemen. I should also like to thank you.
>
> Since being set free, which was only done after I had signed a declaration never to return to Germany, I have been in Belgrade (Yugoslavia) where I was a Professor up to a year ago. I left Belgrade to take up an appointment at the Cancer Institute in Athens which is being set up. As it was not finished at the specified time I first of all went to Vienna. In Vienna the lecturer Dr. Wieser wished to establish with me in the X-ray department of the General Hospital a cancer centre but the Austrian Government would not give me the permit as a German to work. Then I lectured for four weeks in Romania. From Athens I now hear that there has been dispute between the Board of the Cancer Institute and the University does not want any foreigners there and therefore I should not come for the time being. I am now completely

without work and means of earning a living. In Belgrade Austrian Jews are not allowed entry and Germans are expelled after a time. I have been granted a permit to the end of July, that is for two months assuming that it will not be renewed.

In 1933 I was called to Belgrade for 3 years as a professor. That was also why I accepted the call to Athens at the end of 1936 particularly so as my stay was made uncomfortable by propaganda made by Professors Antic and Mahowic against me personally and simultaneously with the rising pro-Nazi movement. During my imprisonment the Yugoslav Minister Lazarevic now in Brussels troubled so much regarding my release and for family in Vienna, that I wish to stress the gratitude I have for Yugoslavia. Lazarevic tried to settle some professors here, including Neumann and myself, which however was not possible owing to the opposition of the doctors here.

I am looking now for a possibility of work and earning a living in any country in the world. I should rather like to spend a term or a year as a guest professor giving lectures in some country on cancer (German or French), and to help organise the fight against cancer. Perhaps the Association here can help. Further particulars about me can be obtained from Professor Dr. Ernst Fränkel.[321]

Adams then responded on 4 July 1938 in his typical firm terms:

Thank you for your letter of the 18th June from Belgrade. We had no share in securing your release from prison in Vienna. I do not know whether this happy result came from efforts made by Dr. Gye or from those made by Professor Fränkel, but whoever was responsible we are delighted that you have secured your release and that you are now outside in the freedom of Yugoslavia. Since we heard from your daughter we have been making enquiries to see if there were any opportunity of finding for you in this country, either research work or invitations to lecture. All our enquiries have so far had no definite result but we will continue to make enquiries and if we hear of any possibility of employment or lecture invitations for you we will inform you immediately.

I should point out that it is a general rule in the Universities and medical schools of this country that the staff retire at the age of 60. Your age therefore will render it practically impossible to find an appointment for you and it would be unwise for you to have any hopes of securing regular employment in the United Kingdom.[321]

Adams also informed Cramer, "We shall do what we can to discourage his coming to Great Britain." Cramer had written to Adams quite negatively on 13 June 1938: "His activities in treating cancer by his remedy make it difficult to justify an attempt to find him a position in this country."[321] The wandering Blumenthal then took up a position in Tirana, Albania, in January 1939 but only stayed for a short period before moving to Tallinn in Estonia in March 1939. He remained there for the start of the war with his wife Elly and grown-up children Zerline and Hilde. They managed to live through the Soviet Union occupation which started in August 1940. When the German army invaded Estonia in June 1941, it was agreed by the Soviet authorities that the whole family could move the long distance to Kazakhstan. They took a train on 5 July 1941 but this was attacked close to Narvas by the German Luftwaffe.[318,320] It seems that none of the family survived this attack. It was a tragic end for this eminent cancer specialist who had extended the lives of so many people through his work in several cities including Berlin and Belgrade.

One of Blumenthal's daughters, Herma, who had written to the AAC with a financial assurance for her father, immigrated to the United States where she lived until 1996. Blumenthal's sister Katharina Buss-Blumenthal was murdered at Auschwitz in August 1942 and his brother Hans, a lawyer, died in the Theresienstadt ghetto in the same year.[322]

In 1943, a concerned Tess Simpson wrote to some of Blumenthal's medical associates asking if anything was known about him. Ernst Fränkel, his former colleague in the Cancer Institute in Berlin, replied on 17 November from Devonshire Place in London:

He left Belgrade for Vienna in 1938, arrested by the Gestapo, set free by some efforts I made together with other English friends, as he wrote to me later. Left for Dorpat in the Baltic, asked me to get him out to England. I tried to get him to South America with the help of Prof. Rosso in Buenos Aires but do not think that he succeeded. I doubt whether he was able to escape into Scandinavia or Soviet Russia when the Germans invaded the Baltic.[321]

Another letter was received by the SPSL from Kurt Stern, who was a student in Berlin in the early 1930s where he had got to know Blumenthal. He wrote from the Polytechnical Institute of Brooklyn on 6 August 1944 with erroneous information:

> The latest I have heard about Professor Ferdinand Blumenthal is that he is alive, contrary to many rumours about his death, and that he is working in Leningrad, U.S.S.R. Since this information emanated from close relatives of Dr. Blumenthal who live in this country, it would seem fairly trustworthy.[321]

This message was, once again, typical of the chaotic situation towards the end of the war when information was scarce and there were many rumours and numerous incorrect communications.

6.5 Hans Hirschfeld

Hans Hirschfeld (1873–1944) was an expert in haematology. He was taken from his home in Berlin to the Theresienstadt concentration camp and died there in August 1944. After the war, a leading member of the

Fig. 6.4   Hans Hirschfeld (~1938).

German community of haematologists who had worked for the Nazis claimed some of Hirschfeld's contributions to medicine as his own.

Hirschfeld was born in Berlin on 20 March 1873 to a Jewish family. He studied medicine at the Friedrich-Wilhelms University in that city and continued to work there on blood diseases. He married Rosa (née Todtmann, born 6 January 1875) and they had two daughters, Inge and Käthe (Kate). As was the case with several of the scientists of Jewish descent discussed in this book, the family officially converted to the Protestant faith. Hirschfeld co-founded the Berlin Haematological Society in 1908.[323] Like Arthur Simons and Ferdinand Blumenthal, he was also appointed to a directorship at the Charité Hospital in Berlin. In 1922, he joined Blumenthal at the Berlin Cancer Institute and was appointed as an Extraordinary Professor. In Berlin, Hirschfeld published over 150 papers on blood diseases and related aspects, including leukaemia and the function of the spleen, and this was exceptional productivity for someone in his field. In 1932, he published the four-volume *Handbook of General Haematology* with the Austrian medical scientist Anton Hittmair and this became widely used.[324]

Following the rise of the Nazis, Hirschfeld was expelled from his professorship in September 1933 together with nearly 300 medical associates of the Charité Hospital. He had written on 24 July 1933 to the AAC stating his hope of being able to undertake medical work in the UK. He received the discouraging response "a medical degree or licence in this country can only be obtained by an examination after one or two year's study at a medical college."[325] His referees included Ferdinand Blumenthal. His papers were sent to William Cramer of the Imperial Cancer Research Fund who was also to comment on Blumenthal in 1938. Cramer responded quite positively on 15 November 1933: "I know by reputation Professor Hirschfeld who is a well-known authority on the physiology and pathology of the blood. I think a place could be found for him, if not in London at least in some provincial hospital." The papers were also sent to the highly influential Professor Archibald V. Hill of University College London who had won the 1922 Nobel Prize for Physiology or Medicine for his work on the mechanics of muscles. He wrote on 8 January 1934 in negative terms:

I know nothing about Hirschfeld myself and his work is entirely out of my own field. Professor Höber does not know him at all, which signifies that he cannot be a person of outstanding distinction. Professor Rosenberg knows about him. He says he is a good man, that he is industrious, that he has produced no really striking ideas or results, that his main work has been on blood pathology — which is rather a different matter from cancer research, that although good he is certainly not in the first class. I should say that this is a man whom we might very justly help if we had plenty of funds available, but if we have, at present, to select cases of outstanding distinction, this is not one of them.[325]

As we have seen in several other applications to the AAC/SPSL, an unenthusiastic report from someone of the influence of A.V. Hill would have meant it was unlikely for Hirschfeld to find a suitable placement in England, despite his impressive publication record. Furthermore, his age in 1934 of 61 was also very much against him as far as the AAC was concerned.

Hirschfeld's application was not then taken further by the AAC for the next four years and he stayed in Berlin. His daughter Ilse, who was also trained as a doctor, met her father for the last time in Vienna in 1936 before moving to Paris and then the USA.[323] The Notgemeinschaft Deutscher Wissenschaftler refugee organisation, at that time based in London, did find a possible place for Hirschfeld in Buenos Aires in Argentina but that was too far for a move.[323]

From 1934, Hirschfeld was able to continue in private medicine in Berlin but this proved impossible after April 1938. He had managed to keep some work going as head of the chemical-serological laboratory in the Jewish Hospital in Berlin. With things getting very difficult, he then applied again to the SPSL on 18 August 1938, saying that it was not now possible for him to continue any medical work in Germany (see Figure 6.4 of a rather pensive Hans Hirschfeld).[325] Tess Simpson replied on 12 December 1938: "Conditions are now extremely difficult. If you have any contacts in the U.S.A., we would advise you to get in contact with them without delay, as conditions in that country are better than in

Europe." The SPSL then received a series of letters about Hirschfeld of increasing urgency. One came from his daughter Kate who was a doctor who had managed to immigrate to Cardiff. Her letter was received by the SPSL on 25 February 1939 and described the very difficult situation for Jewish doctors in Berlin:

> I write about the possibility of getting my father and mother into England as quickly as can be arranged. My father, the haematologist Hans Hirschfeld has already written to you with details of his qualifications. He cannot, however, write to you as freely from Germany as I can, now that I am in England. I beg you to do everything you can to help him, as he is really in danger in Germany. It was only by a lucky chance that he escaped being put into a concentration camp recently from which people, if they come back at all, return ill. We have also recently had a search of our flat by an official who demanded to know when my parents proposed to leave Germany at last.
>
> In October 1938 my parents had to give up their flat as the owners of houses are practically forced by the state to turn out all Jewish doctors. My parents are at present in a small sublet flat, but as the owner is immigrating herself on 1 April, they will be obliged to find another home. This is almost impossible to get now, as Jewish owners are under strict control in their subletting and Aryan landlords are not allowed to make new contracts with Jews. My parents are naturally under great strain not knowing what is going to happen to them or where they can live.
>
> I know you must have lots of applications but I urge you to help my father now in this emergency. He wishes to go on with his research or to carry out research for others and would be glad to have just enough to keep himself and my mother in return for his work. I am hoping very much he will before long get a visa for America where my sister will be within two months (from Paris). But in the meantime, because of the acute mental strain and the real danger, I feel it most urgent to do everything possible to hasten his immigration.[325]

David Cleghorn Thomson's reply from the SPSL on 2 March 1939 was uncompromising:

I am very much afraid that it is quite impossible for this Committee to give a grant in respect of your father. The eminence of his standing has been brought before us and is not in question but most unfortunately it is only possible for us to help with grants in the re-establishment of research workers in academic life, and it is not possible to claim that your father could be re-established in academic life as he is now 66 years old. The only thing which we can do here is to act as trustee for any funds that may be raised to guarantee his maintenance until he can go to America, and to take the necessary steps with the Home Office to secure him a passport when that maintenance has been guaranteed.[325]

The SPSL also had several letters from a close associate of Hirschfeld, Herbert Herxheimer, who had also been a professor at the Charité Hospital in Berlin and had managed to escape from Berlin in 1938 with the help of A.V. Hill. Herxheimer pleaded with Hill to persuade the SPSL to assist Hirschfeld. Hill then wrote to the SPSL about Hirschfeld on 5 April 1939 with some speculative suggestions:

From what Herxheimer says he seems to be rather a distinguished person, but I am afraid he is about 62 years of age, and it goes rather against our principles to try to re-establish him at that age. Perhaps his case could be submitted to some of the members of the medical sub-committee in case they could think of anything that might be done for him. It rather depends on how active he is at 62. Loewi, who is going to India (so I gathered — am I right?) is now, I should think, two or three years older than that. I wonder if there might be a place for Hirschfeld in India. It would not be easy to re-establish him here. Would somebody like him in Ireland?[325]

Here, Hill refers to the Nobel Laureate Otto Loewi who was being supported in 1939, like Schrödinger, by the Francqui Foundation in Belgium. When the war started, Loewi moved to the Nuffield Institute in Oxford and was interned briefly in 1940 before taking a boat to the USA where he had a position at New York University's College of Medicine.[4]

The SPSL also received a letter on 2 August from a close family friend, Lieutenant Colonel Robertson, who stated that, unfortunately, there would be several years before a visa could be granted to enable Hirschfeld to move to the USA but he had managed to raise £150 to assist him and his wife to come to England for a short period. He also suggested contacting Professor William Bulloch of the Lister Institute in London who was familiar with Hirschfeld's eminence in the field of haematology. Tess Simpson then wrote to Bulloch on 3 August 1939 emphasising that the SPSL had been unable to assist Hirschfeld because of his age but asking if he might be able to help him find a position.[325] However, just one month later, the war started and this closed the door on providing assistance for Hirschfeld from the UK.

This was the last communication the SPSL had on Hirschfeld until November 1943 when Tess Simpson wrote to some of his associates in the UK, including Professors Fränkel and Herxheimer. Herxheimer replied, "It looks as though he has been trapped over there," while Fränkel said he had no news and "I was very sorry we did not manage to get this excellent man out on time." He also suggested contacting Hirschfeld's daughter Kate who was previously in Cardiff. She wrote from Birmingham on 27 December 1943:

> In spite of his reputation as scientist, which was so often appreciated by the various organisations with which he and I were in touch, none of them felt able to make arrangements for his coming to this country. In October 1942 he was with my mother deported to an unknown destination but I received a Red Cross message from August 1943 through an uncle who was then still in Berlin and had just received a message that my parents were then alive. The message gave no indication about my parents' whereabouts.[325]

As was the case with several of the scientists discussed in this book who remained in Berlin, including Wreschner, Byk and Lehmann, the assets of Hirschfeld's family were taken by the German State in February 1940.[326] Then, on 30 October 1942, Hirschfeld and his wife Rosa were transported to the Theresienstadt camp on the I/74 "old-age transport."[120] Just one day after deportation, the collector of medical antiquaries Oscar Rothacker entered their apartment to remove valuable books. The possessions from their property in Charlottenburg were sold in January

1943 and their accounts in the Deutsche Bank were confiscated. A report on the Hirschfeld assets was sent to the General Building Inspector for Berlin headed by Albert Speer.[326] It is estimated that over 15,000 apartments and assets of Jewish families were confiscated in this way in Berlin from 1938–1945.[326] After the war, many of Hirschfeld's most valuable items were found to be in the possession of a medical practitioner Dr. Horst Decker and his family.[326]

In Theresienstadt, Hirschfeld, quite remarkably, managed to continue some work in the medical laboratory (see Figure 6.5) and he was even able to give lectures on blood coagulation and formation. In addition, he gave training to some doctors who had also been imprisoned.[326] As we have already discussed, the living conditions at Theresienstadt deteriorated significantly during the war. The Red Cross message indicated correctly that Hirschfeld and his wife were alive in August 1943. He was still giving

Fig. 6.5   Hans Hirschfeld in Theresienstadt, 4 August 1943. Sketch by Max Plaček (1902–1944).

lectures, together with other colleagues at Theresienstadt, in July 1944 just one month before he died there on 26 August 1944 at the age of 71.[326] His wife Rosa managed to survive the war and went to stay with their daughter Kate in England. Rosa had become very weak after being incarcerated in Theresienstadt for nearly three years and died in 1948. From England, complicated efforts were initiated to claim reparations for what was taken from the Hirschfeld's during the war.[326] His daughter Ilse had a full medical career in New Jersey and specialised in paediatrics before retiring in 1975.

After the war, the haematology community in Germany actively ignored Hirschfeld's contributions to their field despite his international reputation. It was only in 2012 that detailed information on Hirschfeld's work was published by the German Society for Haematology and Oncology.[326] The second edition of the *Handbook of General Haematology* had been published in 1957 by the publisher of Hirschfeld's original Handbook, Urban and Schwarzenberg, in the name of his previous co-worker Anton Hittmair along with Ludwig Heilmeyer. Hirschfeld, however, was not mentioned (see Figure 6.6).[327] Hittmair himself had been deported to the Flossenbürg concentration camp in 1939 but had been released after he assisted in containing an outbreak of dysentery in the camp.[326] After the war, he eventually became the vice-president of the European Haematology Society. Heilmeyer was a Luftwaffe doctor during the war and was responsible for a hospital containing Soviet prisoners of war where over 150,000 died due to lack of medical care. He also supported Wilhelm Beiglböck in the post-war Nuremberg trial of doctors who had been involved in Nazi experiments on people. Beiglböck was found guilty and, after his release, Heilmeyer employed him in Freiburg.[328] Heilmeyer became widely known and honoured as a leader in Germany in the field of haematology. He became the founding rector of the new Ulm University in 1967.[329]

However, Heilmeyer was found out in due course.[329] The city of Ulm removed Heilmeyer's name from a street.[330] The same was done in Freiburg where the Ludwig-Heilmeyer-Weg was renamed the George-de-Hevesy-Weg. In addition, the University of Jena, where Heilmeyer received his habilitation, removed a plaque that had been erected in his name.[328]

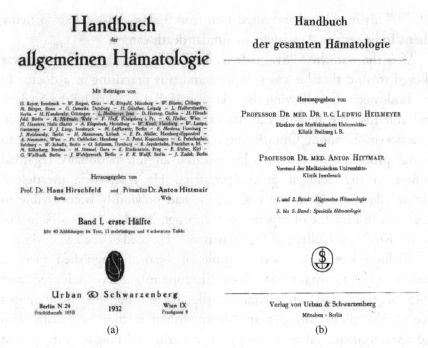

Fig. 6.6   (a) Front page of *Handbuch der allgemeinen Hämatologie*, edited by Hans Hirschfeld and Anton Hittmair, 1932 (left). (b) Front page of *Handbuch der gesamten Hämatologie*, edited by Ludwig Heilmeyer and Anton Hittmair, 1957 (right).

In 2021, the city of Ulm in Germany dedicated a "Hans Hirschfeld Platz" which is connected to the Albert Einstein Allee and James Franck Ring.[331]

Not all the distinguished medical scientists who had been unsuccessful in their applications to the AAC/SPSL and were sent to Theresienstadt perished during the war. Vojtech Adalbert Kral is one example whose case is similar to that of the mathematician Paul Funk from Prague. Kral was born on 5 February 1903 in Prague and graduated from the German University there in 1927. He was appointed as a Professor of Psychiatry in Prague until 1938 when he was dismissed at the end of the year. He applied to the SPSL on 15 June 1939 but, as we have seen in other applications from Prague, this was much too late for the SPSL to take any action.[332] He was sent to Theresienstadt on 22 December 1942 together with his wife. He gave expert advice to the camp administrators on an epidemic of encephalitis that might have spread from the internees to the staff.[333] This may have prevented him from being chosen, like so many others, to be sent to Auschwitz or similar death camps.

After the war, the SPSL made its usual enquiries about Kral and was relieved to find that he was still alive and was practising as a doctor in Czechoslovakia.[332] Following the takeover of that country by the Soviets, Kral and his wife fled to Canada. There, he built a leading clinic in geriatrics in London, Ontario, and published many papers linked to mental problems of old age. He became an internationally regarded pioneer in the field of psycho-geriatrics. He received a prestigious acknowledgement from the UK, which had previously been unable to assist him in his time of urgent need, through his election as a Fellow of the British Royal College of Psychiatrists.[333] He died at the age of 85.

Adalbert Kral is a good example of what distinguished medical scientists from Germany and Central Europe might have achieved after 1945 if they had not been murdered in the war. It remains incomprehensible that so many distinguished medical scientists, who had saved numerous lives in cities such as Berlin and Prague in the period before World War II, were wiped out by the Nazis. The ability of their vicious regime to provide medical care for their citizens was diminished considerably as a result.

CHAPTER SEVEN

Refugees in Biology

In the 1930s, Biology as an academic subject did not have as many scholars as chemistry or physics in Germany and other countries in Central Europe. The number of applicants to the AAC/SPSL classified as biology was 103, whereas the numbers were 176 and 160 in chemistry and physics, respectively.[334] Furthermore, some biomedical scientists (including the Nobelists Krebs and Chain) would probably have also classified themselves as biologists or biochemists, such is the overlap between biochemistry, zoology and medicine.[21] The biological scientists formed a diverse group.[335] Furthermore, it is clear from the cases examined that refugee biologists were somewhat more flexible in the countries they could move to for their research, with South America and Asia being quite popular destinations with interesting possibilities for field studies. However, there were a small number of biologists who applied to the AAC who are relevant to the current study. Two who had particularly adventurous lives stand out — Hellmuth Simons and Vladimir Tchernavin.

7.1 Hellmuth Simons

Like Fritz Houtermans, Hellmuth Simons (1893–1969) was a larger-than-life character who had a hugely adventurous career. He did important research in developing tests for use in pathology and bacteriology. In the 1930s, he managed to come to England for a short period but then went briefly to Russia and, after that, Paris. In the Second World War, he was arrested more than once in France and was imprisoned and tortured. He had a charmed life, and managed to escape to

Fig. 7.1   Hellmuth Simons, March 1943.

Switzerland where he continued his scientific research. On the Nazis top-wanted list, he was also involved in spreading influential rumours that Germany had a biological weapons programme.

Simons was born in Düsseldorf on 17 April 1893 to a Jewish family. His elder brother was the medical scientist Arthur Simons whose tragic life has already been discussed in this book. In Düsseldorf, Hellmuth Simons attended the Reform-Realgymnasium and then studied zoology, physics and chemistry in Freiburg. He also worked briefly at the Zoological Station at Villefranche. In his own words, he expressed to the AAC:

> During the war I was attached to the Pathological Institute of the Academy of Medicine at Düsseldorf training a number of laboratory assistants in histological microscopical methods and diagnosing parasitological material. I also did some work on fixing agents and began my work on nagana — which was the subject of my thesis and which led to my appointment as malaria expert for the 7th Army Corps ... In 1917 I attended Küster's classes for microbe culture methods.[336]

After World War I, he went to Würzburg to complete his studies in zoology and biology and then returned to the Medical School in

Düsseldorf to continue research. In 1922, he published a paper on novel rod-shaped bacteria which were subsequently given the name *Simonsiella*.[337] He also worked at the Hygienic Institute in Cologne to become acquainted with medical-bacteriological methods. Due to the financial difficulties in Germany after World War I, Simons became a chemist in a private laboratory combating insect pests to add to his income. He also carried out a research project for the German Government on gnats and mosquitos at the Hinsbeck Limnological Station. From 1932–1933, he was attached to the German Patent Office giving advice on food chemistry. All of these activities gave him very broad interests which he would put to good use in his subsequent career.

Following the rise of the Nazis, Simons applied to the AAC in December 1933.[336] His referees included Albert Einstein (even though Simons had no interests in physics) and Chaim Weizmann. On 19 September 1933, Einstein wrote:

> Dr. H. Simons tells me that he, like so many of my unfortunate colleagues, is trying to find in England or through English help, a new field of activity. I know him personally as a man almost fanatically devoted to science. I have before me his testimonials from bacteriologists which show that he is an able and reliable research worker. I should be glad to know that he obtained the possibility of devoting his powers to the service of a good cause.[336]

For his other referees, Simons also named the eminent British biologists David Keilin of the Molteno Institute and Magdalene College in Cambridge and Julian Huxley of the Zoological Institute at King's College London. He was able to visit the laboratories of his referees shortly before he applied to the AAC. On 29 November 1933, Keilin wrote, "From my knowledge of his published work and conversations with him I have formed a high opinion of his general knowledge of biology and his versatility."[336] Keilin had become a world-leading authority on entomology and parasitology, and had also invented the word "cytochrome" for the redox-active proteins which are a vital component of energy transfer in cells.[338] Julian Huxley, who was the brother of the writer Aldous Huxley, wrote on 22 December 1933: "Dr. Simons is a very able scientist both in pure and applied work; his work appears to be both versatile, accurate and important. His ability and obvious energy would clearly make him a most useful worker in any laboratory concerned with parasitology or with applied work on cereals."[336] Although Simons did not come from a major academic appointment in Germany, he had several publications that clearly had impressed these leading biologists in the UK.

Simons was an effective networker and arranged to see William Rintoul, director of the Imperial Chemical Industries, who had been allocating several grants to German scientific refugees to come to work in UK laboratories. He also visited Professor J. W. Munro at Imperial College London. Munro then wrote to the AAC on 11 June 1934 to say that he was hoping for a grant from ICI for Simons to work on the infestation of grain and flour by insects. However, he stated that competition for resources in this area was scarce, despite the clear applications. Simons, however, did start to work with Munro but without payment.

There was another venerable British institution that had a very keen interest in Simons. This was the Security Services MI5.[339] Simons was staying in London with a close friend Otto Lehmann-Russbüldt who was a pacifist from Germany who had already written on the threat of German rearmament. He had been arrested by the Nazis after the Reichstag fire but had managed to escape to the Netherlands and then on to London. His activist associates were known to MI5 and that had put him under suspicion. The MI5 interest had been heightened by an article in 1934 by Henry Wickham-Steed, a former editor of *The Times*, entitled "Aerial Warfare: Secret German Plans" which also mentioned German research on airborne biological warfare.[340] As noted in the correspondence of MI5 with Balmoral Castle, this sensational article had even caught the interest of King George V as the suggestion had been made that biological weapons could be released in the London and Paris underground rail systems.[339] As a journalist, Wickham-Steed had declined to name his sources for his sensational information and MI5 had strong suspicions of both Lehmann-Russbüldt and Simons together with a mysterious Dr. Karl Meister. One MI5 report stated:

Dr. MEISTER was at one time in partnership with Dr. Rudolf SIMONS. They owned jointly a chemical Factory in Saxony. Owing to the nature of the work carried out, the factory was uninsurable and when an explosion occurred, they lost a great deal of money. Dr. SIMONS is a Jew and was at one time very wealthy. He has studied Zoology and Biology and is described as being a typical "professor". Owing to Nazi persecution Dr. SIMONS left Düsseldorf either at the end of August or beginning of September 1933 and came to UK, where he has rich relatives.[339]

MI5 always wrote surnames in capitals in all their reports and Rudolf refers to the final Christian name of Hellmuth C. R. Simons. Lehmann-Russbüldt produced a book with the title *Hitler's Wings of Death* which exposed secret German aircraft production in contravention of the Treaty of Versailles.[341] A second book was also eventually produced in 1937 by the German pacifist Heinz Liepmann entitled *Death from the Skies — A Study of Gas and Microbial Warfare* and Simons was mentioned in the book as having provided key information.[342] On 2 February 1936, MI5 intercepted a letter from Liepmann to Simons on the research being done for his book and gave a summary:

> Has been dealing now for over a year with English and American publishers about the gas-book and has worked at it, but until now has had very little satisfaction. Now he has sold his new book "… will be punished by death" to an English publisher and has also attracted his interest to the gas-book. The writer is apparently writing a book on chemistry and wishes the addressee's help. The latter evidently does not wish his name to be mentioned in connection with the book. He must give a source for his information however. He told the addressee not to make any different remark to anyone else and then they can avoid all risks. Will probably have a decision from the English publishers and from America in a week or two. Keen to complete the book now at all costs as it is becoming acute. Needs the addressee's help and asks him to ring up and arrange a meeting. As he is going to have a share in the profits, he must help with the costs. If they cooperate well they will have the work finished in a few weeks. Awaiting his telephone call.[339]

An analysis after the war did not find evidence of a major biological warfare research programme in Germany but there was an intensive chemical weapons programme, derived from I.G. Farben research on insecticides, an area in which Simons was expert.[10,193] This research produced the deadly chemical nerve agents tabun, in 1936, and sarin in 1938. Most fortunately, Hitler did not authorise the use of these most dangerous of chemical compounds in the war despite the urging of some of his advisors.

Simons' visit to ICI had also seemed unusual when he had asked some probing questions on its production of explosives. This was reported to MI5 and heightened their suspicions even further. After this, his actions were followed and his mail was opened. One of the typical reports from MI5 stated:

> In December 1935 information was received that Major Lefebure of I.C.I had been approached by two German refugees named Otto LEHMANN-RUSSBUELDT and Hellmuth SIMONS in connection with a book which they had written on armament manufacture. Major Lefebure originally got to know them through the publisher Mr. Stanley Unwin, who asked his advice on an earlier book of theirs on German aircraft.
>
> In the course of a meeting which took place in December, LEHMANN and SIMONS put to Major Lefebure a number of leading questions about armament production in this country and made some rather informative statements about German manufacture of poison gas. It was also stated that SIMONS had an appointment with Lord Melchett on 18th December. Major Lefebure afterwards informed the Chemical Defence Research Department of this conversation and asked whether there was any reason to distrust LEHMANN and SIMONS. He also asked whether any attempt would be made to investigate the information which they had indicated they possessed. Major Lefebure further told the story to Mr. F.A. Freeth of I.C.I. who passed it on to the police. Any offer of secret information on the part of LEHMANN and SIMONS is open to some suspicion as they are known to associate with persons suspected to be connected with a foreign intelligence service, and it is more than possible that the intention behind their remarks to Major Lefebure was to extract information rather than to impart it.[339]

A subsequent related MI5 report indicated that Simons and his friend were considered more to be a nuisance than dangerous:

> A caller at New Scotland Yard on 13th December said that both these refugees are writing a book on "armaments", and that a member of the staff of the Imperial Chemical Industries from whom they attempted to obtain information considered them "either fools or knaves".
>
> It is true they are writing a book on "armaments". LEHMANN-RUSSBUELDT is doing the drafting and SIMONS is procuring the data from books at the British Museum and other places. These men are ardent pacifists, and both were forced to leave Germany because of their political opinions. They are writing books here, not only to air their ideals, but to gain a living. LEHMANN-RUSSBUELDT, in particular, is considered by the Germans to be a dangerous political opponent to the Nazi regime, and all his writings are directed against it.
>
> They both have a persecution complex; see spies everywhere, and are constantly warning fellow refugees to be on their guard. As far as is known they have not taken part in political propaganda inimical to the interests of this country.[339]

Possibly because of the MI5 interest in his discussions with ICI, which was communicated to ICI, no funding came through to Simons from that source for his work with Professor Munro at Imperial College. This lack of funding was a concern for the London-based Professional Committee for German Jewish Referees who wrote to the AAC about the matter on 9 September 1934. However, on 21 October, Simons wrote to the AAC in a slightly irritated tone:

> I am very happy to tell you that since the 8th of October I have found an appointment as assistant bacteriologist with Dr. Weizmann's Laboratory in London. The Jewish Academic Committee has given me a grant of 150 £ for nine months. When Dr. Weizmann will be content with my work — and just have had success in certain investigations — he will take me over to his plants in Paris, Barcelona or perhaps to the new Agricultural Institute in Rehoboth (Palestine). Now you can cancel my name from your list and destroy the copies of all my testimonials if you like.[336]

However, this happy situation in the laboratory of the future President of Israel did not last long. On 15 January 1935, Simons telephoned the AAC to say that Weizmann was closing his London laboratory and he had not been paid while he had been working there.[336] He was hoping that his grant from the Jewish Academic Committee could continue. In the meantime, he was looking into the possibility of immigrating to the Soviet Union, with encouragement from David Keilin. The US Emergency Committee in aid of Displaced Foreign Scholars had also received a letter from Simons and asked the AAC for advice. The response from Walter Adams on 26 April 1935 was:

> We have consulted several people about the prospects of Professor Simons and it is fairly clear that there is practically no chance of his obtaining an academic post within Great Britain and we have advised him, therefore, to try to find an industrial position, and both Mr. Makower and ourselves have recommended his name several times for industrial openings, but so far without success. He has, for some time, been working in London in the Chemistry Laboratory of Dr. Weizmann, although without any payment from Dr. Weizmann himself. He has so far been supported by the Jewish Professional Committee and Mr. Makower tells me that they have supported him chiefly because they thought that there might be a slight hope of his finding a position with Dr. Weizmann in Palestine.[336]

His grant did enable Simons to stay in the UK for a few more months. He was able to visit the pharmaceutical company Hoffmann La Roche in Basel, Switzerland, in July 1936. This visit was noted by the German SS and the company was also a very useful contact for Simons when he managed to move to Switzerland from France in 1943.[343]

Simons had been advised by Keilin in Cambridge to consider a position in Russia. As we have seen, several leading refugee scientists in the UK had recommended Russian laboratories including Born (for Alfred Byk) and Peierls (for Fritz Duschinsky). However, they were all to come to regret this advice after the purges in the Soviet Union in the 1930s. Simons then wrote to Professor Yevgeny Pawlovsky of the Institute of Experimental Medicine in Leningrad who was an expert on parasitology. His letter to an intermediary in Paris was intercepted on 17 February 1936 by MI5:

> May I take the liberty of asking whether I might not be able to render you useful, scientific work to your Institute. My special subject has been for a long time the study of parasites, particularly research into the struggle against malaria. I am a student of Professor DOFLEIN (formerly of Freiburg) in Breslau and from 1915 to 1919 I was expert in malaria with the VII Army Corps and was entrusted with special orders from the higher sanitary officers. As a malaria specialist, I took a special course with Professor WASIELEWSKI (then in Heidelberg) and studied medicine for several years, supervised closely dozens of malaria courses at a malaria hospital and took a special course in microbiology with Professor KUESTER in Bonn. Among others I have published various works on malaria which were highly esteemed by the medical profession, and which H. ZIEMANN in particular in "Mense's Handbook of Tropical Illnesses" has appreciated in detail. I enclose a list of my publications. I would be able to place at the disposal of your institute valuable optical instruments, my collection of microscopic preparations and my countless books as well as many special editions. I am 42 years old. You will be able to find out about me in detail from Professor D.W. YAKIMOFF (Leningrad) and Professor D. KEILIN (Cambridge).[339]

A reply from a Dr. Olshwunger in Paris on 18 February 1936 about the move of Simons to Russia was also intercepted by MI5. Its secretive tone was of interest although at that time the Soviet Union was not considered such a threat as Germany:

> They are in receipt of his letter of the 17th. They inform him that on his arrival at the frontier he will find a letter containing money at the station post-office. This money will cover his travelling expenses to Moscow. The cost of the journey to the place of work will be paid by the government. The state pays the family's travelling expenses from the frontier, independent of the time of his journey ... The testimony of EINSTEIN will probably not help him much, nor harm him.[339]

Simons worked at the Tropical Institute in Leningrad from 1 March 1937 to the end of June 1937. He took his son Werner with him who was now aged 16. His daughter Gerda remained as a student in Cambridge where an aunt also resided. He had recently divorced his wife Gertrud who remarried and went to live in Lima, Peru.

At this time, the atmosphere in Russia was changing very quickly, as we have seen from the cases of Duschinsky, Wasser and Houtermans. After two months, Simons' pay cheques in Leningrad stopped.[339] In June 1937, Marshal Mikhail Tukhachevsky was tried for being a German agent. This placed German émigrés in Russia under suspicion and some scientists were arrested. Simons was in contact with the Institut Pasteur in Paris and he managed to move there by travelling first from Russia to Denmark on 26 June 1937 and then to France.[339] This clever biological scientist even managed to persuade the Russian government to pay for his trip, which was something he also did with the Swiss government when he was making arrangements to immigrate to the USA in the 1950s. When Simons went to work in the USA in the 1950s, his short visit to Leningrad caught up with him when there were questions raised by the US authorities about possible communist links.[344]

On 2 February 1938, the SPSL were informed, "He left Russia some months ago and is now in Paris." Simons then wrote to the SPSL somewhat formally on 30 August 1938:

> The Rockefeller Foundation in Paris advised me to write to you and I am taking the liberty to do so. I am a German refugee and have been living in Paris since July 1937 and before that in London for 3 and 1/2 years. My age is 45, I am married and have 2 children under age. I am a doctor of natural science in protozoology, mainly in relation to tropical medicine, and can produce excellent references as a research worker. I have specialized in the subjects of parasitology and I have applied to you because all my endeavours to obtain permission to work either in France or in the French colonies have failed, although I have recently been able to publish some papers in French scientific journals. I should like to ask you whether you can help me to get any post in England, the English colonies or anywhere else.[336]

His letter was passed on to Mr. Makower, Chairman of the Professional Committee for Jewish German refugees, whose reply on 2 September 1938 had a distinctly negative tone:

> I do not think that he has acted in a straightforward way with us. He was maintained by us for over three years, provided with a ticket to U.S.S.R. and since then has been quite out of touch with us and only by chance we heard that he had returned from Russia. We should be unwilling to take any further interest in his case and I doubt whether you would have much satisfaction if you took up the case.[336]

Tess Simpson took the hint and replied to Simons on 3 September 1938: "I am afraid however that the prospects at the present time are not at all good." Simons was always persistent and he wrote again to the SPSL from Paris on 7 October to say that his daughter had informed him of a public appointment for an Entomological Officer for Research at the UK Agricultural Research Board. He asked if the SPSL could "recommend him warmly for this post." Tess Simpson, however, replied at once to say the SPSL could not intervene in such appointments.[336] That was the last communication with Simons from the SPSL until enquiries started to be made about his whereabouts after the war.

Simons had managed to continue research in the Institut Pasteur in Paris until the start of the war. Just two days after the announcement of the war, he was interned as a German national in the camp of Colombes, near Paris, on 5 September 1939. He was then caught up in a series of extraordinary and dangerous incidents. He wrote a vivid eye-witness account of his movements over the next three years that illustrates the chaos in France following the defeat in 1940 to Germany and a series of remarkable escapes.[345] He had taken his son Werner to France who had volunteered to join the Foreign Legion. On several occasions, this enabled Hellmuth Simons to be released from captivity in France.

Simons was set free from internment on 1 December 1939. He was not allowed to continue working at the Institut Pasteur but was able to use the library there. However, following the attack by the German military forces on France, Simons was interned again by the French authorities on 13 May 1940 in the camp at La Braconne, north of Bordeaux. Just 24 hours before the Nazis were due to arrive there, the internees were released on 21 June at 11 pm after setting fire to the huts where they were living. Simons managed to trek across the countryside to Marseilles where he was able to stay in the St. Jean Fortress that had been converted into a training centre for the Foreign Legion.[345]

Before the war in Paris, Simons was financially supporting himself by translating into English reports from French journals, theses and books of biology, medicine and linked subjects. These works were then forwarded to the University of Pennsylvania in the USA, who then distributed the documents as *Biological Abstracts*. While in Marseilles, he managed to continue this activity despite all the challenges of the war.[13] He even asked the US editor of the *Biological Abstracts*, John Flynn, to write to the US consul based in Marseilles to support the application of a visa to go to the USA.[13,345] In this process, Flynn received assistance from Alvin Johnson, the influential director of the New School for Social Research. Johnson also managed to get Gerda, Simons' daughter who had moved to the UK, released from internment on the Isle of Man.[346]

Eventually, Simons was able to secure a berth on a ship from Lisbon to New York but this was scheduled only for April 1943. Before this date, he was arrested in Marseilles on 26 August 1942 and sent to the Camp-des-Milles internment camp near Aix-en-Provence, where Wasser and Dreyfuss had also been sent. Following the occupation of Vichy France, in 1942 by German troops, transports were arranged to take Jewish people in Les Milles to camps in the north of France, such as Drancy, from where they were transported to death camps. Simons, however, was spared this transportation due to, once again, his son's membership of the Foreign Legion.[345] He was then sent to yet another camp, this time at Rivesaltes near Perpignan but was released from there by a kindly camp commander who was secretly a member of the French Resistance. Simons had used his chemical expertise to help produce forged documents for endangered internees including some politicians and this assisted his release.

Unknown to Simons, the Nazis had established a Wanted List of nearly 3,000 important people who they thought were working in the UK and who should be arrested as a matter of priority by the SS during the expected invasion of Britain. The list included leading politicians such as Winston Churchill and Charles De Gaulle. Also included was "Doctor Hellmuth Simons, Emigree, believed location London, married with 2 children, 1917 assistant at the pathological inst. Düsseldorf, 1920-5 researcher at Biebrich-Amoneburg Laboratory, 1925-33 Researcher in Germany, 1935 researcher at Hoffmann-La Roche Basle, emigrated to London in 1936. Speciality parasitology, microbiology, milling chemistry. Wanted by the Amtsgruppe."[343] It is ironic that Simons had been arrested several times in France and no connection was made with the SS Wanted List.

After a short period in Marseilles, Simons was arrested again following the shooting of a German officer, and was tortured and incarcerated in a pitch-dark prison. He was informed that he was going to be shot as a reprisal for the shooting along with 19 other hostages. However, the French Resistance, who were aware of his work with forged documents, freed him and three others from the prison before the shootings took place.[345] The Resistance also arranged for money and papers which enabled Simons to take a train on 6 March 1943 to La Roche sur Foron-Monnetier near the Swiss border. From there, he safely crossed into Switzerland in the plain vertically below the summit of Mont Salève and spent the night in a farm. He then gave himself up to a local farmer who contacted the Swiss police, and he was transferred to a refugee camp in Val Fleuri in Verbier.[344]

At the refugee camp, Simons was interviewed by the immigration police (see Figure 7.1). His persuasive abilities again came to the fore as, not being a minor like Elisabeth Wasser, he had to convince the authorities that if he was returned to France he would at once be arrested and, if he was allowed to stay in Switzerland, he would not be financially dependent on the State. He informed the Swiss authorities:

> I was arrested as a hostage in Marseille on January 26, 1943 and was informed that the Germans would take revenge on March 8 for an allegedly unknown perpetrator killing some German soldiers. I was informed I would be shot. At the time of my arrest, I was abused with rubber clubs together with other hostages (all intellectuals from different

nations and denominations). Following a bribe by some French friends I was able to run away on 6 March.[344(t)]

He also repeatedly informed the Swiss interrogators that he was working on *Biographical Abstracts* for the University of Pennsylvania in Philadelphia that was bringing him a regular income. He gave the names of several friends who could vouch for him including a cousin who was acting as the Consul for Honduras in Berne. He had to inform the immigration authorities of the full details of his family. He was divorced from his wife Gertrud Bella Simons (née Hess and born on 8 February 1897) who was now living in Lima, Peru, where she was working as a laboratory assistant in metal chemistry. He also stated that his daughter Gerda Ellen Clara (born on 19 May 1919) was living at 4 Mortimer Road in Cambridge, England, while his son Werner Heinz Bernard Simons (born 17 May 1920) was still in the Foreign Legion in Algeria. In addition, he declared that his brother Arthur Simons had been deported from Berlin to Poland. He also stated that he had been given an American entry visa in 1942 that lasted until 5 February 1943 but he had been unable to use it because visas for Jews were blocked in France on 21 July 1942 when the Gestapo initiated major roundups of Jewish people.[344]

In addition, he was able to emphasise to the Swiss authorities, who were well aware of the importance of the pharmaceutical industry in their country, his previous scientific visit to Hoffmann La Roche in Basel in July 1936 when he gave advice on his ideas for biological testing. He had also visited Switzerland as a tourist several times in the 1920s and was very familiar with the country. After being held in the Charmilles and Aeugstertal reception camps, he was moved to an internment house in Magliaso on Lake Lugano and then to Welahem, near Zurich.[344]

The contrast between a safe life in Switzerland and the dangers in France in 1943 could not have been greater. Compared to Emanuel Wasser, who was also in the Camp-des-Milles, subsequently arrested by the Gestapo in Clermont-Ferrand and then quickly sent to Auschwitz via Drancy, Simons was most fortunate to be alive. The leading physiologist from the University of Berne, Professor Alexander von Muralt had collaborated with the Nobel Laureate Otto Meyerhof at the

Kaiser-Wilhelm-Institut for Medical Research in Heidelberg in Germany in the early 1930s and there he had come across Simons. He was to set up the Swiss National Science Foundation in 1952. Von Muralt was a significant supporter of Simons in Switzerland and helped him find a placement at the prestigious ETH in Zurich where he was able to continue his work for *Biological Abstracts* through his translation of Swiss biological and medical works.[344] His translations were then sent via diplomatic mail in Berne to the University of Pennsylvania. This work provided a regular income in Switzerland for Simons which was supplemented by money from aid agencies. Simons also informed the authorities that he was cataloguing the *Scientific Reports of the Royal Society of London* from 1800–1915 for use in Switzerland.[344]

Von Muralt even applied for funds for Simons from the Swiss Department of the Interior with the clever argument that Simons' translation work was communicating Swiss scientific findings to the outside world. The appearance of Simons in Switzerland was also noticed by Hoffmann La Roche who invited him on 1 June 1944 to again visit its laboratories in Basel, where he had visited briefly in 1936, and submit a proposal for a research collaboration in the biomedical area.[344]

Simons' claims that had been published in England suggesting that the Germans had a significant biological warfare research programme also appeared again in Switzerland. Allen Dulles, Head of the US Office of Strategic Services in Berne, who was to become Director of the CIA, sent the following intelligence telegram to Washington on 8 December 1943:

Formerly employed in bacteriology research in connection with its warfare manifestations, Prof of Biology Hellmuth Simons, a German Jewish refugee, who worked at I.G. Farben, now works at the Zurich Polytechnic Institute. He feels that bacteriological warfare toxin "bacillus botulinus" will be the German secret weapon. According to Simons the toxins can be readily produced in quantity in such plants as the following: I.G. Farben at Hoechst, the research laboratory in Berlin of Heereswaffenamt Amt HardenBergstrasse, or the Behring works at Marburg. Prof. Simons says he has worked at the British Museum Library and at Cambridge's Molteno Institute, where Wickham-Steed and Prof. Keilin were both very familiar with him. Prof. Simons appears to have reached his conclusions through deductive evidence rather than from exact new evidence from Germany. Please inform us if this matter is sufficiently interesting to you to mean my following up the preliminary memorandum he has given me.[347]

The rumour even reached General Eisenhower who was making plans for the D-Day invasion of France in June 1944. There was then some talk of developing a possible antidote to the toxin or inoculating in advance the troops who would take part in the June 6 invasion although no such action was taken.[348] As was also discussed earlier, after the war, it was determined that the Germans did not have a biological warfare initiative.[349]

While in Switzerland, Simons also had a romantic relationship with the Swiss painter and sculptor Alis Guggenheim with whom he had a prolific and personal correspondence (see the photograph of Figure 7.2).[350] He wrote to her with concern that the "yankees" were destroying his hometown of Düsseldorf with bombs. Their relationship did not seem to continue after the war. Alis Guggenheim provided financial support for Elisabeth Wasser to stay in Switzerland (see the chapter on Emanuel Wasser).

Simons was now able to send letters from Switzerland to his daughter and aunt in Cambridge, providing they were written in English, and receive letters in return. He was relieved to hear that his son Werner had

Fig. 7.2   Hellmuth Simons with Alis Guggenheim in Switzerland, 1943.

Fig. 7.3   Hellmuth Simons, 1945.

survived the war in North Africa where the French Foreign Legion had been based. His letters, as previously, were intercepted by MI5 who were surprised to find he was now in Switzerland.[339] MI5 reported on 1 March 1944: "He was arrested twice in France, presumably on account of the French authorities' dislike of refugees in general and Communists in particular."[339] It seems that MI5 were more worried about Communists than Nazis, even in 1944. MI5 were also baffled by a letter to his daughter of 22 July 1944 in which Hellmuth Simons described life on Lago di Muzzano with Alis Guggenheim as "living like a God" where he was able to make necklaces from the shells of plants "of great rarity," feast on "fat eels and tenches" and enjoy the "bee-hives, tomatoes, plums, grapes, apricots and apples."[339] This "paradise" was quite a contrast to the austerity in England.

By the end of the war, Simons was in somewhat better shape than when he first arrived in Switzerland (see the photograph in Figure 7.3 above). The institutions in Switzerland had noticed the importance of his research work. On 31 October 1946, the following press release, entitled "Rapid Detection of Blood Infections" was published by the Swiss Society for Research Expeditions, based in Basel:

After first painstaking investigations, which required thousands of experiments, thanks to the courtesy of the university institutes and the cantonal hospital in Zurich, Professor Simons discovered in 1945 the Thedan blue dye which can detect smallest amounts of the causative agent of malaria. Also, the trypanosomal diseases (such as grinding disease, Chagas disease, tsetse disease in Africa) and sptrochata diseases, syphilis, severe infectious disease such as jaundice. The Thedan test allows for important microscopic examinations which used to take 254 hours and now can be done in 10 seconds in even primitive working conditions. The optical properties of the dye transform the blue colour of the parasites in the bright field of the microscope into a faint red in the dark field. Prof. Simons is busy with the practical evaluation of other important findings in the field of pathogenic mutants and our secretariat will continue to inform the public about this work.[351(t)]

Then, the following announcement was made in the international journal *Science* on 15 November 1946:

H. C. R. Simons, known for his research work on pathogenic blood protozoa, is now in Switzerland in relatively good health, according to Walter Pilnik, Hammerstrasse 20, Zürich 8, Switzerland. In 1933, Prof. Simons, unable to return to his house in Düsseldorf, because of the Gestapo, fled to England, where he continued his research work in London and Cambridge. In 1937 he went to the Institut Pasteur, where he invented his technique in diagnosing blood in a protozoa by saponine-methylene blue and taurocholate of methylene blue. In 1943 he was arrested by the Nazis but was liberated 48 hours before his execution by French partisans and was able to reach Switzerland. He now continues his abstracting work for *Biological Abstracts*, begun in France, and has improved his saponine-methylene blue method under unfavourable conditions, since Swiss regulations make it difficult for foreigners to accept paid work.[352]

After the end of the war, Ilse Ursell made her usual enquiries on those scholars who had applied but had then lost contact with the Society. On 16 April 1945, the UK Research Bureau wrote to say that it had information that a Hellmuth Simons was living in Magliaso, Switzerland, and "it was just possible that this is the same person."[336] The SPSL also wrote on 4 May 1946 to David Keilin at Magdalene College, Cambridge, to say it had been informed by a Red Cross Agency that Simons had survived the war and were wondering if Keilin had an address. Keilin replied with the address: Pension Plattenberg, 8, Schönleinstrasse, Zurich. Ilse Ursell then wrote to Simons to say that the SPSL had not heard from him since 1938.[336]

Simons had made quite a name for himself with the pharmaceutical industry in Switzerland and he continued to collaborate there after the war and published papers in the general area of bacteriology under the name "H.C.R. Simons." His link with the University of Pennsylvania through *Biological Abstracts* remained strong. On 19 August 1947, there was a report in the Lebanese (Pennsylvania) Daily News with the headline "Refugee Scientist is Studying Polio," mentioning a move to the USA:

A new field for research is open to Dr. Hellmuth Simons, 53-year-old German refugee scientist, who is resuming his quest for a possible new treatment for poliomyelitis at the University of Pennsylvania. Ousted from the University of Düsseldorf by the Nazis, Dr. Simons escaped wartime Germany into France and, subsequently, Switzerland. He later conducted research in London, Paris, Moscow and Zurich. Dr. Simons is the discoverer of theodane blue, a stain used by biologists to identify blood parasites under a microscope. His work in the United States will include a staff appointment to *Biological Abstracts*, published at Pennsylvania by an independent organization.[353]

However, he made arrangements with the Swiss authorities for permission to return to Switzerland from time to time. The persuasive Simons had also asked the generous Swiss to provide him with a loan to cover the costs of his move to the USA (as he had even done in Russia some ten years before), although a few years later they asked him to return the loan.[344] He crossed the Atlantic several times in the 1950s, declaring himself as "stateless" to the immigration authorities at Ellis Island.[58]

Then, on 7 September 1952 from 205 North 36 Street, Philadelphia, USA, Simons wrote an urgent letter to the Swiss Immigration Office:

Hereby I allow myself to approach you with the request to provide information about me during my stay in Switzerland from March 1943 until my departure for the USA at the end of April 1947. This information is now very urgently needed for my forthcoming naturalization as an American citizen. I fall under the law of Senator Pat McCarran (Nevada) because in March 1937, on the urgent advice and with the resources of the Jewish Refugees Committee (London), I went to the Soviet Union with my son, who was still a minor at the time, where I had been offered a senior position as a parasitologist in the Caucasus. Luckily the Russians sent me and my son from Moscow to the Institut Pasteur in France for further special research.

For a few weeks now, the "McCarran Act" has been asking all persons who were in the Soviet Union and who have applied for American citizenship to provide police proof of where they have been since leaving the country. This is why I contact you directly with my concerns. I hereby ask the police department to issue me a double certificate of good conduct in German for the duration of my stay in Switzerland ...

Unfortunately, in 1937 I was persuaded by a very important English colleague and British employer to accept a research position in the Soviet Union. I now fall under the very strict American laws on naturalisation. I have to prove that this stay in Russia was purely for economic purposes, based on English documents. However, according to American law, I still require proof that I have not been a communist since my return to Europe and that I have not had the slightest suspicion of flirting with the communists. Switzerland was the last country I stayed in before leaving for the USA, which is why a letter from the police department is of the greatest value to me.[344(t)]

Clearly, his short visit to Leningrad in 1937 was becoming a problem with the authorities in the USA, who had become highly suspicious in the 1950s of any communist links. An FBI report of 14 August 1952 on Simons, which was passed on to MI5, stated in somewhat sensational terms:

In 1937, at the suggestion of numerous English scientists, he went to Russia with his 16-year old son and there became head of a malaria research team. After several months, he was transferred to the Pasteur Institute in Paris ... He stated that he was employed by the British

Government during World War II as an expert in bacteriological warfare and claims to be responsible for deterring a bacteriological invasion of England.[339]

Around this time, his son Werner Simons, who was essentially stateless, had applied for naturalisation in Britain. However, this was initially denied due to the MI5 "communist" suspicions remaining on his father and also because Werner had joined his father when he was 16 years old on their brief visit to Russia in 1937. An MI5 report on Werner's application, with reference to his father, stated on 21 October 1951, "It would appear from what we know of his character and actions before the war, that Dr. SIMONS was a rather unbalanced, excitable and not very reputable individual."[339]

The well-organised Swiss authorities duly obliged in sending the documents Hellmuth Simons needed which they had readily available in their archives (and still do nearly 70 years later).[344] However, as was the case with Paul Dreyfuss, Simons' stay in the USA did not last very long and on 17 May 1955 he wrote once again to the immigration authorities in Berne asking for permanent residence in Switzerland.[344] He stated that he would receive an income from his private research and was also receiving a reparations pension from Germany and a small pension from the USA. He continued to undertake research back in Switzerland and his last biomedical paper was published in 1958 in German in the journal *Deutsche Medizinische Wochenschrift* on "Is Disseminated Sclerosis caused by Spirochaetes" and giving addresses for correspondence in both New York and Basel.[354]

Hellmuth Simons, who had such a colourful, adventurous and productive life, died on 19 March 1969 in Berne at the age of 76. His son Werner, who had indirectly saved his father's life in France more than once by joining the Foreign Legion in 1940, was eventually able to live in Cambridge in England, presumably to be close to his sister Gerda. He died there in 1994 at the age of 74.[355]

7.2 Vladimir Tchernavin

Vladimir Tchernavin (1887–1949) was unique. He was not a refugee from the Nazis but purely from Soviet Russia and he was not from a

Fig. 7.4   Vladimir and Tatiana Tchernavin, 1933.

Jewish family. He was an expert on the biology of the salmon. Following the purges in the USSR, he had a remarkable escape from the Soviet Gulag with his wife and son. The SPSL went out of its way to help this unusual zoologist come and work in England but he died tragically not long after the end of the war.

Tchernavin was born in 1887 in Tsarskoye Selo, near what was then called St. Petersburg (and was named Leningrad after the revolution, and more recently has reverted to the older name). He came from a noble family that had limited wealth. He was married to Tatiana and their son Andrei was born in 1918. She was the daughter of a popular professor Vasily Sapozhnikov (1861–1924), a botanist who was a pioneering explorer of the Altai Mountains where Russia, China, Mongolia and Kazakhstan converge. Tchernavin also took part in numerous scientific expeditions which included Western Mongolia, the Volga, the Caspian Sea, Tian-Shan, Lapland, the White Sea and the Arctic Ocean.[356] In 1914, he was conscripted into the Russian army but was wounded and released. In 1916, he obtained a B.Sc. degree in natural sciences from Petersburg University. He then went into fisheries research which was his lifelong interest. He took a leading part in an expedition by the Department of Agriculture collecting material on the Salmonidae family of fish. His master's thesis at the Petersburg Agricultural Academy was on "The origin of the secondary sexual characters in Salmonidae."

From 1918–1921, after the Revolution, there was a great famine in Russia but Tatiana had a fascinating job cataloguing the treasures and archives in the recently vacated Imperial Palaces first at Pavlovsk and then Peterhof near St. Petersburg.[357] She was involved with a team that restored and opened for the first time ten palaces and turned them into magnificent museums, which told the story of Russian life and culture over the previous two hundred years. Tatiana also had some income from writing fairy tales for children. In due course, she became the assistant curator of the Hermitage Museum in the applied arts section.[357] This experience would be useful later when she was able to lecture on the great Imperial Russian treasures in the UK.

Vladimir Tchernavin became a senior reader in zoology at the Agricultural Institute from 1921–1923 in St. Petersburg and then was appointed Professor of Icthyology. Due to his expertise, he was appointed in charge of Fisheries Legislation at the Board of Fisheries in Moscow for 1924–1925. From 1925–1930, he was head of the scientific laboratories of the Northern Fisheries Trust in Murmansk. His work was consistently interrupted by the OGPU, the Soviet Secret Police, which eventually became organised under the notorious NKVD. Due to the desperate famines in the Soviet Union, administrators involved with food supplies were always under suspicion. The first warning of trouble for Tchernavin was when his letters from Murmansk to Leningrad began to arrive many days late. This was well known as a sign that the dreaded OGPU was intercepting them.[358] Shortly after this, Tchernavin was interrogated in Murmansk, but he was not arrested initially. What followed was a remarkable story of escape that was subsequently told in detail by Tatiana Tchernavin.[357]

In September 1930, the Soviet newspaper Pravda had the headline "Discovery of a Counter-Revolutionary Organisation to Wreck the Workers' Food Supplies."[358] There followed a vicious purge of scientists and officials involved with meat, fisheries and agriculture supply. In Tchernavin's own words, "I was arrested in 1930 and in April 1931 without a trial was sentenced by the OGPU to five years penal servitude at the Solovetski concentration camp on the White Sea ... All the leading specialists in fisheries in the USSR were shot by the Bolsheviks (I got only

5 years hard labour in the Prison Camp)."[358] His wife was also arrested for a short period. Their son delivered food parcels to their prisons which kept them alive and he had to sell some of the family furniture to pay for the food parcels.

The island camp of Solovetski had been set up "for opponents of the Bolshevik Revolution" in a remote location. There, Tchernavin was able to work as a fisherman and a fishing instructor, although the living facilities were appalling. This useful work allowed him to be transferred to Karelia, close to Finland, to assist with the fishing industry. He even did some primitive research and found a new way to kill salmon that his captors appreciated. At this camp, his wife Tatiana and teenage son Andrei were able to visit him.[357] He told them he was developing a plan for escape from the Soviet Union with them. He would mention to her the code word "north" in a letter when things were ready.[357] They were to bring a razor for him to shave and fresh clothes or else he would be spotted. A compass was also essential.

Tatiana then sold all of their belongings to provide some money for their escape. She also made a last visit to the Hermitage and was sad to find that many great paintings, including Titians, Botticellis, Rembrandts and Van Dykes, had been sold. Then, Tatiana and her son were allowed to visit Vladimir again in August 1932. Her husband had found a rowing boat in which the family escaped. The two adults and their son, now aged 14, then had to walk for 22 days through forests and swamps. They found berries and mushrooms to eat. Eventually, they came across forest workers in the territory of Finland and were saved.[357] Very few prisoners escaped from the camps in the Gulag, but the determined Tchernavin was one and it was even more remarkable that he managed to leave with his family.

On 22 March 1934, the AAC was surprised to receive a letter from a Russian lady unknown to it by the name of Tatiana Tchernavin:

> I wonder if you would be inclined to consider the case of my husband, Tchernavin. He was professor of marine zoology in Petersburg and did a great deal to organise the fishing industry under the Soviet Government. In spite of his being devoted to his work he was arrested in 1930 as a "wrecker" and after several months in prison was sentenced to five years

penal servitude in the Solovetski concentration camp. In the summer of 1932, he succeeded in escaping from there, with me and our little son, and we are now living in Finland. At the moment I am on a visit to London and would be glad to give you all the particulars you may require about my husband's academic qualifications.[356]

The numerous letters in the AAC/SPSL archives now tell, very largely, the remaining story of the extraordinary life of the Tchernavins. Tatiana was sent the usual AAC form for her husband to complete which he did after he also arrived in London in 1934. The AAC took great care with this unique application from a Russian scientist who they did not know and had no links with Germany. The tragedy of the Soviet Gulag was only a mild rumour at that time and it was only after refugees such as Vladimir and Tatiana Tchernavin published their stories that the horrors were revealed (see Fig. 7.4).[357,358]

The application to the AAC from an expert in the biology of the salmon was unique. He described himself as an "outlaw in the USSR." His referees were Finnish and Polish fish biologists who were unknown in England. After his arrival in London, he had made contact with the natural history section of the British Museum. His referees spoke warmly about his expertise in fish zoology and fisheries. Professor Siedluki from Krakow in Poland said, "You have an eminent and reliable scientific worker, worth of employment in any scientific institution connected with fishery." M. Zarotschenzeff, from the American Z Corporation, Rochester, New York, wrote, "I am acquainted with his scientific and practical works in the fish industry. His work merits much attention." As was usual, opinions were requested from experts within the UK. John Norman, who was curator of the natural history part of the British Museum, and three years before had written *History of Fishes*, was quite encouraging. He wrote on 18 May 1934:

I am afraid that I know very little of Professor Tchernavin, and of his published works only two or three have found their way into our library. I have turned up his paper on "The nuptial changes of the skeleton of the Salmon", published in 1918, and have gone over this with Dr. Regan, who has done a good deal of work on the Salmonidae. We conclude that this is a sound piece of work, and that, if his other work on the group is of a similar quality, he should be worthy of encouragement. I don't think that there is anybody else who could give you further information about him, and it is my impression that his work is very little known outside Russia. This is doubtless due to the somewhat obscure publications in which his papers have appeared, and to the fact that only one of them has found its way into the Zoological Record.[356]

Boris Uvarov, who was also from the British Museum (National History), wrote a well-informed letter:

I knew Tchernavin in the Petrograd University where we have been working in the Zoology Department at the same time, and later, shortly before the war, when he was working in the Zoological Museum of the Russian Academy of Sciences, where I also used to work. His later career is also known to me from independent sources and I will be able to confirm his statements regarding it. I know that he has always been regarded by his senior colleagues as an outstanding worker in his own branch. His principal papers are known to me, and although I am unable to judge their merits in his special field, I can see that they are obviously of a high quality and contain valuable general biological ideas.

Knowing Tchernavin's work in the past I believe that he could be most usefully employed in a big Museum, where his thorough knowledge of the systematics of fishes could find application. I have discussed his case unofficially with the Director and the Keeper of Zoology Department at this Museum, and they believe that Tchernavin's services may be of definite value to the Museum. A working place and an abundance of material could be placed at his disposal, but there is no chance of the funds being available for remuneration. If, however, the funds can be provided by the Council, the Museum authorities would be glad to accommodate him.[356]

Uvarov had been educated in St. Petersburg, like Tchernavin, and was a world-leading expert on the biology, ecology and control of locusts. He had been recruited by the British Museum in 1920. He was to be elected a Fellow of the Royal Society in 1950 and was knighted (KCMG) in 1961 for his direction of the Anti-Locust Research Centre in London. This was a remarkable achievement for a Russian emigrant.[359]

Harold Munro Fox of the Natural History Section of the British Museum, who was also shortly to be elected a Fellow of the Royal Society for his zoological research, was similarly positive in his letter of 14 June 1934:

> Unfortunately, I am personally not acquainted with Professor Tchernavin's work since the whole, or practically the whole, of it has been published in Russian and so far as I know has not been translated into English. I recently, however, had an opportunity of consulting one or two continental scientists in fishery matters on the subject and they informed me that Professor Tchernavin's work was altogether excellent. They also stated that he had not received full credit for some of the work that he did in the North of Russia and that quite recently this has been taken by other people.[356]

With these strong references, a successful case was made to the AAC for a significant grant of £250 per annum for Tchernavin. The AAC then wrote on 23 March 1935 to the very-well-connected Julian Huxley, who had also been consulted about Hellmuth Simons:

> I see that after all Chaplin is not coming to London so that our only hope is that Mr. H.G. Wells has made a serious attempt to interest him in the work of the A.A.C. Perhaps Chaplin may be coming to Europe later and if Mr. Wells has not been successful another attempt might be made. I shall of course inform you as soon as I hear anything from Mr. Wells.
>
> Would you allow me to give a letter of introduction to you to Professor Vladimir Tchernavin, a Russian refugee, who has been maintained by my Council for the past two years. He has been continuing his research on the problems of the sex of salmon and has now almost completed this particular study for publication. He tells me that his conclusions differ very widely from those of Dr. Tate Regan and this of course raises a rather difficult problem because Professor Tchernavin has been a guest at the Natural History Museum and has profited from their hospitality.[356]

The entrepreneurial efforts of the AAC to pique the interests of the world-famous actor and film producer Charlie Chaplin and the prolific writer H.G. Wells are clear from this letter, which also raised a difficulty that was emerging about Tchernavin's research work. In the meantime, he had been working unpaid at the British Museum (National History) with John Norman who had become much more enthusiastic than before. He now wrote on 10 September 1935:

> He has been engaged in studying the cranial osteology of the Salmonoid fishes, especially salmon and trout. His investigations have been undertaken with a view to determining the changes which take place in the skull and jaw during growth and at the breeding season, the differences to be found in the two sexes, and also to try and discover a satisfactory basis for the classification of these fishes. Professor Tchernavin has made considerable progress with his work and has prepared a very valuable series of specimens. I regard his research as of first-class importance and look forward with interest to its completion. His knowledge of this particular group of fishes seems to be extensive and as far as I am able to judge he is a careful and accurate worker.[356]

On 17 April 1936, a notice was placed by Lord Rutherford in *Science* announcing that the Academic Assistance Council was to be replaced by the Society for the Protection of Science and Learning:

> The Academic Assistance Council was formed in May, 1933, to assist scholars and scientists who, on grounds of religion, race or opinion, were unable to continue their work in their own country. Its services have been needed chiefly to help the 1,300 university teachers displaced in Germany, but it has also assisted refugee scholars from Russia, Portugal and other countries ... The council hoped that its work might be required for only a temporary period, but is now convinced that there is need for a permanent body to assist scholars who are victims of political and religious persecutions. The devastation of the German universities still continues; not only university teachers of Jewish descent, but many others who are regarded as "politically unreliable" are being prevented from making their contribution to the common cause of scholarship.
>
> The council has decided to establish as its permanent successor a Society for the Protection of Science and Learning, which will continue the council's various forms of assistance to scholars of any country who, on grounds of religion, race or opinion, are unable to carry on the scientific work for which they are qualified. One function of the society will be

to build up an Academic Assistance Fund to award research fellowships, tenable in the universities of Great Britain and other countries.[360]

A second letter was also published in the same issue of *Science* and also simultaneously in *The Times*:

Sir, In your issue of March 18 you published an account of the new plans of the Academic Assistance Council for the reorganization of its work and the creation of a permanent body — the Society for the Protection of Science and Learning with the general aim of safeguarding the freedom of learning. The undersigned, and with them their friends, collaborators and pupils, feel that they should not let this moment go by without publicly expressing their gratitude to the Academic Assistance Council, as the executor of the good will and friendship of their English friends.

Hundreds of scholars, faced with the necessity of abandoning their studies, have sought and found advice and help from this organization. It is due to the devotion and energy of the members of the council that difficulties which at first appeared insurmountable have been overcome, and that the council, in collaboration with other organizations, has succeeded in placing 363 out of 700 displaced scholars. In reality, far more has been achieved than these numbers indicate. It is in the very nature of the problem that even where no material assistance was possible, help could be given by satisfying spiritual needs. The warm sympathy extended to all who approached the Academic Assistance Council has helped in hundreds of cases — this part of its work cannot be illustrated in figures.

The Academic Assistance Council is coming to an end in its emergency form, but we and our friends will endeavour to make it remain unforgotten. May we hope that the continuation of our scientific work — helped in no small measure by its activities — will be an expression of our gratitude?

Albert Einstein, E. Schrödinger, V. Tchernavin.[361]

It is likely that the second letter was organised by the Council of the AAC and probably by Lord Rutherford himself. The choice of Einstein as a signatory was presumably made because he was the world's most famous living scientist and was also a Jewish refugee from Germany. Schrödinger may have been asked to sign because he was a rare example of a "politically unreliable" scientist (in the words of the Nazis) who was not Jewish but had left Germany after Hitler came to power. Tchernavin, who did not have the fame of these two Nobel Laureates, was also presumably chosen because he was a unique example of a scientist who had escaped from the Gulag in Russia. He was in excellent scientific company with his signature to this letter.

The courageous Tatiana Tchernavin had begun to make a name for herself in London partly through her unique museum work in Russia in the Imperial Collection. She had also published her book *Escape from the Soviets* in 1934 which described her extraordinary experience and that of her husband, who were very rare escapees from the Soviet Gulag.[357] In the next year, Vladimir Tchernavin also published his own story *I Speak for the Silent* (1935) which emphasised the attack of the Soviets on scientists.[358] With her husband struggling to find a secure employment, Tatiana was interested in using her own talents and experience to find a position in the USA connected to museums. She contacted the AAC for advice on an ambitious lecture tour she was planning to give in the USA and Canada. Adams of the AAC then wrote to William G. Constable on 6 November 1935:

> Mrs. Tchernavin, the wife of our one Russian grantee, Professor Vladimir Tchernavin the zoologist, was formerly assistant curator at the Hermitage decorative arts section. As I told you, she proposes to make a visit to the U.S.A. and Canada soon after Christmas to give popular public lectures on her experiences in Russia and hopes to take an opportunity while there of getting into contact with museum and art gallery people. She had wide experience in popular educational work in museums in Russia and would like to see if she could find similar work in the United States. Her itinerary includes Baltimore, Toronto, Kansas City, San Francisco, Santa Barbara, Pasadena, Louisville and Philadelphia. The final details of her lecture tour are not fixed so that she might include other cities on her tour. If you could suggest names of persons whom she ought to try to see while in the United States she would be very grateful.[356]

Constable was director of the newly formed Courtauld Institute of Art. A distant relative of the famous painter John Constable, he was soon to be appointed as director of the Boston Museum of Fine Arts. Despite Tatiana Tchernavin's unique experience at the Hermitage, one of the world's leading museum and art collections, the art and intellectual establishment in London was not enthusiastic. Sir Bernard Pares, a historian of Russian History, was an exception and arranged for her to give a lecture in London. From Boston, Constable gave a negative view to the SPSL on 1 June 1938 about a possible position for her in the Frick Library in New York:

> As regards Mrs. Tchernavin, I am sorry to say I cannot speak with any conviction on her behalf. I thought she was a nice woman and certainly very knowledgeable, but I did not find that she had the kind of qualifications which are essential in such a library as that of Miss Frick. As a matter of fact, her name has already come up for consideration here, details concerning her having been sent to Paul Sachs. The best thing I can suggest is that she gives my name as a reference to Miss Ethelwyn Manning, who is the Librarian of the Frick Library, and I can then tell her what I think ...
>
> My only impression is that Mrs. Tchernavin's best chance is in some kind of school work. I said this to Paul Sachs, and he has let it be known in several quarters.[356]

Paul Sachs was one of the founders of the Museum of Modern Art in New York. During the war, he was involved in setting up the military task force which saved many valuable art works in Europe. He was also a partner of the investment bank Goldman Sachs.[362]

The AAC/SPSL tried very hard indeed to find a secure position for the unusual Vladimir Tchernavin. Possibilities of positions in Portugal, Michigan and Aberdeen were followed up as were attempts to find support from the fisheries industry but nothing came through. Walter Adams wrote to Uvarov with concern on 7 January 1937: "I am very worried about the present position and future prospects of Mr. Vladimir Tchernavin ... He is so modest himself that I am afraid he will take little action in his own interests."[356] Then, a position at Edinburgh University was found for him in October 1937 researching on the salmon but only at the level of a research student.

During the war, Tchernavin continued to research and publish papers in the *Proceedings of the Zoological Society of London* on the breeding characteristics of the salmon and related topics.[363] He became more dependent financially on his wife and, when the war started, she obtained a position at the UK Ministry of Information as a translator. Tatiana Tchernavin even provided subtitles for the patriotic film *In Which We Serve* in Russian, which was directed by Noel Coward and David Lean and included famous actors such as Richard Attenborough and John Mills. It was distributed worldwide and was nominated for an Academy Award.[364] It is ironic that her work helped to make this film available for an audience in Russia, the country which had treated her and her family so badly.

After the war, the Tchernavins continued to live in England. Vladimir published a paper in *Nature* in 1946 from the Department of Zoology at the British Museum (Natural History). It had the intriguing title "A Living Bony Fish which Differs Substantially from all Living and Fossil Osteichthyes."[365] He also wrote to the SPSL on 14 January 1946 about Professor T. Spiczacow from the University of Cracow who had been one of his original referees. He stated that Spiczacow had been detained in a camp for displaced persons and wondered if the SPSL could help bring him to England in the same way as he had made his way there.[356] There is no record of any assistance being provided for Spiczacow. It has to be said that the AAC/SPSL went out of its way to find a suitable position in fisheries research for Tchernavin himself, and never really succeeded.

Then, on 5 April 1949, the Inquest section of *The Times* reported under the heading "Professor's Death":

> Inquests were held at Hammersmith yesterday on Professor V. V. Tchernavin, 62, a distinguished biologist and a refugee from Russia since 1932, who was found dead in a flat at Emperor's Gate, Kensington, and on Miss Anna Drew, 40, the tenant of the flat, who was found dead there the previous night. Evidence was given that Miss Drew's death was caused by a growth and a verdict of "death from natural causes was recorded". A verdict that Professor Tchernavin "took his life when the balance of his mind was disturbed" was recorded, and the Coroner

added: "He looked after Miss Drew with great devotion and as a result of her death he took his life."[366]

It was a true tragedy that, after all Tchernavin had been through with his family, he should die in this way. His mind must have been severely disturbed by the shock of Anna Drew's death. This notice was the last one kept in his extensive AAC/SPSL file.[356] A more detailed obituary was also published in *Nature* where it was stated, "He was a man willing to sacrifice more for his ideals of liberty and truth than most of us."[367] He was a rare example of a Russian refugee in Britain.[368]

Tatiana Tchernavin stayed in England after the death of her husband. She was granted British nationality in 1950. She died on 1 March 1971 at the age of 83. Their son Andrei, who had accompanied his parents across the wastes of north Russia to escape to Finland at the age of 12, was made a British national in 1949. He qualified as an engineer and had a successful career in this field in England. He took part in a BBC Eyewitness Film called *Gulag* in which he revisited the St. Petersburg flat where his family lived, and the camp where he and his mother visited his imprisoned father and took a rowing boat to freedom.[369] He died in Oxfordshire in 2007. He was buried in St. Mary's Churchyard in South Perrott, West Dorset, alongside his mother and father. Their gravestone reads, "Escaped Stalin's Terror."

CHAPTER EIGHT

Refugees in Engineering

In the 1930s, engineering was a very practical subject with many companies and industries in Germany and elsewhere developing new applications and products. As is the case today, some engineers took up university positions after experience in the industrial world. Compared to the pure scientists who have already been discussed, refugee engineers were more flexible in finding suitable employment in companies in safe countries where there was a reasonably buoyant market for their expertise. Consequently, the number of engineers, 54, who applied to the AAC/SPSL was smaller than in physics, where as many as 160 applied. We could only identify one applicant in engineering who failed to get assistance in the UK and also had a tragic end.

8.1 Alfred Rheinheimer

Alfred Rheinheimer (1884–1944) was an engineer from Bavaria who worked in Glasgow before the First World War when he was interned in the Isle of Man. In 1938, he was also interned for a brief period in the Sachsenhausen concentration camp.

Rheinheimer was born on 26 June 1884 in Thaleischweiler in Bavaria. He came from a Jewish family but his wife Helene was not Jewish.[370] They had no children. He graduated in 1907 from the mechanical engineering department at the Technical University in Munich. He then took employment in countries outside of Germany on "calculations and design of machines, lifting hoists and conveyer plants" in Budapest, Paris and

Fig. 8.1   Alfred Rheinheimer in his AAC application (1933).

Glasgow (with Sir William Arrol & Co.).[371] His international positions had made him fluent in English and French.

With the start of the First World War and being placed in Glasgow, he was interned from 1914–1919 in the Knockaloe Camp, Isle of Man.[372] Following the war, he moved back to Germany working for engineering companies in machine fabrication. In 1926, he turned to academia and took up a post as a lecturer in mechanical engineering at the Technische Staatslehranstalten in Hamburg. He taught students in the design, drawing and manufacture of engines and machine parts.

On 28 June 1933, Rheinheimer was informed of his dismissal from his lecturing post following the implementation of the new laws for the civil service. Within one week, he had applied to the AAC (see Figure 8.1).[371] He stated that he would consider a lecturing position and also a commercial post that used his engineering expertise and experience. His referees included Professor Siefken, Director of the Hamburg Technische Staatslehranstalten, and also some representatives of engineering companies. As was the usual procedure with the AAC, his application was sent to a leading university expert in the field of the applicant, in this case Cecil Lander, head of the Department of Mechanical Engineering at Imperial College in London. Rheinheimer was able to make a short visit to London, and Lander responded to the AAC on 13 November 1933:

> From November 1926 to July 1933 Dr. Rheinheimer was a lecturer in mechanical engineering and hoist engines in the Technische Staatslehranstalten, Hamburg. This is a technical college of good standing, but not of University rank. Dr. Rheinheimer appears to have been the lecturer in charge of the subject under the direction of Professor Siefken, the director of the college. Most of his life has apparently been spent in either the design department or the selling department of mechanical engineering firms. From March 1913 to September 1914, he was engaged in designing cranes and structures at Sir William Arrol and Co., Parkhead, Glasgow, and, if necessary, further information could be obtained from Mr. R. G. Shepherd, who was his chief at this firm. Dr. Rheinheimer showed me a letter which he had received from this firm in August 1914, stating that if Scotland Yard offered no objection they would be willing to re-engage him, so that presumably he gave satisfaction. From what I could gather, Dr. Rheinheimer is a sound mechanical, engineer, but has by no means reached the top ranks of his profession.[371]

As we have seen in other applications to the AAC, a lukewarm response from one of the UK academic experts meant that priority would not have been given to the application, even though Rheinheimer had engineering experience in Scotland. The AAC had a well-informed network in the academic world, but this did not extend to the industrial sphere. A position in Bolivia was suggested by the AAC and Rheinheimer responded that "he was not entirely unwilling" to take up a position there but his clear preference was for employment in the UK.[371] It seems he did not take up the opportunity of being employed again in Glasgow, a possibility that was implied in the letter from Professor Lander.

With his application to the AAC not successful, Rheinheimer stayed at Hohenzollernring 101, Altona-Ottensen, a suburb of Hamburg.[370] He obtained employment as a teacher at a Jewish school. Following the

Kristallnacht of 9 November 1938, he was arrested and was sent to the Sachsenhausen concentration camp. He was released on 21 December 1938. He hoped that his wife being non-Jewish would protect him but this proved not to be the case. He was eventually denounced and arrested in 1944. The records show that Alfred Rheinheimer was murdered at Auschwitz on 16 August 1944.[370]

On 8 March 1946, the SPSL received a letter from the UK Search Bureau for German, Austrian and Stateless Persons from Central Europe about Alfred Rheinheimer. It stated, "No trace could be found of the missing person. The following offices could give no information: German Police, Labour Office, Foreigner Labour Office, Criminal Police, Card-Index for people who were killed in air-raid, Hauptstandesamt, Jewish organisation."[371]

Rheinheimer has been commemorated recently in the Stolpersteine project. This is a ten-centimetre brass "stumbling block" which has been placed outside the last known residence of a victim of the Nazis. Alfred Rheinheimer's Stolpersteine is placed at Hohenzollernring 101 in Hamburg.[370] Several other victims discussed in this book have also had Stolpersteines placed in their memory, including Erich Aschenheim (Hindenburgstrasse 49, Remscheid), Paul Eppstein (Ludwigkirchstrasse 10, Berlin), Hans Hirschfeld (Droysenstrasse 18, Berlin-Charlottenburg), Robert Remak (Manteuffelstrasse 23, Berlin) and Arthur Simons (Kurfürstenstrasse 50, Berlin).[373]

CHAPTER NINE

Refugees in Social Sciences

It is of interest to examine if there were any major differences between applications to the AAC/SPSL in sciences and social sciences (but we have not analysed the large number of applications in the humanities). Law and economics stand out as subjects of social sciences in which the universities in Europe were highly active. In both law and in finance, the refugees proved to be fairly flexible as there were also non-academic opportunities in other countries.[374] In contrast to Germany and Austria, social sciences was not a well-developed field in UK universities before World War II and there were very few opportunities for positions. Consequently, it was quite common to bypass the assistance agencies and directly contact relevant companies and institutions. This flexibility meant that there seemed to be smaller numbers of refugee lawyers and economists who applied to the AAC/SPSL and had tragic ends in the war compared to those we have discussed in pure sciences, although it must be emphasised that the Nazis were ruthless in their treatment of Jewish lawyers who failed to escape.[375]

Many refugee social scientists also found positions in the USA. The New School for Social Research in New York went out of its way to find research positions for refugee scholars in the social sciences and the humanities. Accordingly, it was often dubbed "The University in Exile." Several refugees who came to the School received support from the Rockefeller Foundation, and an American Council for Émigrés in the Professions was set up. As opposed to organisations in the UK and elsewhere in Europe, the School often became the first point of call for applications from social scientists in difficulty in Europe in the 1930s.[376]

Nevertheless, the AAC and SPSL did still continue to receive some unusual applications in the social sciences. In 1934, Professor Ludwig Waldecker applied to the AAC.[377] He had been a Professor of Law in Breslau and then Cologne. The trusted expert in law from whom the AAC often obtained advice was Albrecht Mendelssohn Bartholdy. He was the grandson of the composer Mendelssohn and had been a leading authority in Germany on Anglo-Saxon Law. Bartholdy founded the Hamburg Institute for Foreign Policy and was one of the lawyers who represented Germany in the Paris Treaty negotiations after World War I. He was expelled from his Professorship in Law in Hamburg in January 1934 and, with the assistance of the AAC, had taken up a Fellowship at Balliol College, Oxford.[378] Bartholdy described Waldecker on 7 December 1934 as "not only an excellent scholar but an independent man of progressive views."[377] However, Ernst J. Cohn was another eminent jurist and a Professor of Law in Cologne who had been helped to immigrate to England by the AAC.[379] On 28 January 1934, he had stated, "Waldecker was and is a Nazi."[377] With a recommendation such as this, the AAC would not be providing assistance.

Some 14 years later, on 16 January 1948, Ilse Ursell wrote again to Ernst Cohn and also included a personal message:

> I have found in the file of Professor Ludwig Waldecker, originally of Breslau and then of Cologne University, a note with comments from you saying that the man was a Nazi and had only been transferred from Breslau to Cologne. According to a report which appeared in *The Times* of 11th March 1936 he was dismissed in Cologne ... I should be interested to know whether you have any further comments.
>
> You might be interested to know that a student of Cologne University, whose parents were friends of my own, recently sent me an enthusiastic report about your lectures there.[377]

Cohn replied on 19 January 1948 again in less than complementary terms:

> I have to inform you that Professor Waldecker died some two years ago. He was certainly a Nazi and boasted to me personally of it. However, he

was a quarrelsome and incompetent man and may have fallen out with
the Nazis as he fell out with everybody else.[377]

Another unusual refugee who started off as a social scientist and then
wrote with distinction on a huge variety of topics is the brilliant polymath
Robert Eisler.

9.1 Robert Eisler

Robert Eisler (1882–1949) was an Austrian intellectual and scholar who
had very wide-ranging interests, from the economy of banking to ancient
cosmology and the historical picture of Jesus. Born into a rich family, he
was detained in Dachau and Buchenwald in the late 1930s. After his
release, he spent his last years with limited resources in Oxford.

Eisler was born in Vienna on 27 April 1882 to a Jewish family. His
father Friedrich was a successful partner in a manufacturing company and
his mother came from a wealthy family that had made its money investing
in the American railroads. Initially, this financial advantage allowed
Robert Eisler to move between a remarkable range of intellectual studies
without holding down a permanent appointment.[380,381] He studied at the

Fig. 9.1   Robert Eisler (1928).

University of Vienna where he obtained his first doctorate in economics. At the age of only 20, he published five essays on Studies in Value Theory.[382] Following a tour of European museums and galleries, he turned his interest to art history and Greek philosophy, and obtained a second doctorate at the Art History Institute in Vienna. He had something of an inexplicable aberration in 1907 when, on an impulse, he stole a precious codex from the Archbishop's Palace in Udine. He was arrested and self-harmed in prison.[381] He was released provided he spent time in a sanatorium. He then married Rosalia ("Lili") von Pausinger in 1908, a Baroness from Austria and daughter of the landscape painter Franz von Pausinger, and he became a Roman Catholic.

In World War I, Eisler, somewhat surprisingly, served with distinction. He was an infantry officer and was made a knight of both the Order of Franz Joseph and the Iron Cross. In the next few years, he interacted with intellectuals in many fields and wrote several books with varied titles such as *Money: Its Historical Origin and Social Significance* and *The Enigma of the Fourth Gospel*.[383,384] In the autumn of 1922, Eisler visited Oxford and a note written in 1949 gives a clue as to why this turned out to be such an important place for him:

> On an unforgettable sunny day 37 years ago I visited Oxford for the first time as a young graduate of another university, now in ruins, almost as old and equally famous but — as most universities on the continent — without the autonomous, corporate, intellectual life of resident scholars which has survived in Oxford's picturesque old colleges uninterrupted from the Middle Ages. My host, a great European and famous historian now departed, showed me the Martyrs' Memorial and I was told in moving words: "This ... is not a red brick university of yesterday and tomorrow. The whole place where thou standest is holy. Here men have died for their convictions." I have never forgotten this introduction to the glories of Oxford.[385]

In 1925, Eisler took up a post with the International Commission on Intellectual Cooperation of the League of Nations in Paris. He had support from both Albert Einstein and Sir Gilbert Murray, the Regius

Professor of Greek at Oxford University, in taking up this post. He had sent one of his books on finance to Einstein who replied on 31 January 1925: "The little book about money appealed to me particularly for its erudition, intelligence and sentiment ... I am persuaded you are a very worthwhile candidate, and you can be sure that I would most warmly support you."[386] However, the Austrian Government objected to his appointment, with his arrest for the art theft in Italy being a major factor, and he was removed from the post. Einstein wrote to the Austrian historian Alfred Pribram on this matter on 1 May 1926:

> Through Gilbert Murray (Oxford) and myself, the archaeologist Dr. Robert Eisler was appointed to the Institut International de Coopération Intellectuale in Paris. We did this because of the generally recognized scientific merits of this man. But he bears the odium of youthful misconduct (he took home, somewhere in Italy, a document for photographing without permission, but with no intention of misappropriation).[387(t)]

Eisler then continued to write on the New Testament and the Early Church (see Figure 9.1).[381] He also lectured at the Sorbonne. With the economic depression after the Wall Street Crash in 1929 affecting the financial situation of his family, Eisler's extraordinary mind turned back to economics. He wrote popular books on how he thought the banking system should be reformed including the titles *The Money Maze* and *Stable Money*.[388,389] He lectured at Rhodes House in Oxford in 1932 on his proposal for the British Empire to have a common currency, an idea which links to the euro which was adopted by many European countries seventy years later. He was even invited to lecture on this topic to the Parliamentary Finance Committee at the House of Commons and to the Foreign Exchange Commission in Washington, DC.[381]

The new developments in Germany in 1933 did not seem to affect Eisler's life significantly until the Anschluss in Austria in 1938. His Oxford links and his concern that the Nazis were rapidly controlling the whole of Europe persuaded him to apply to the SPSL in March 1938.[390]

He listed an extraordinary set of referees, including Einstein, Gilbert Murray and Kenneth Clark, the Director of the National Gallery. Then, on 20 July 1938, David Cleghorn Thomson wrote to a friend, Baroness Anna von Schey-Karomla, about a worrying rumour:

> Strangely enough one of the very first things I hear is a rumour that charming little Dr. Eisler, whom I met several times with you, is in a concentration camp in Austria. I am writing off at once to you to ask whether you have any direct information regarding this as I should dearly like to help this organisation to help him if it is possible. Could Inge write back by return if you have any detailed information about Dr. Eisler which might help us should this rumour be true.[390]

Thomson then wrote again on 23 August 1938 with some good and bad news:

> I still alas have no information about Robert Eisler, the last being that he has actually been chosen and appointed to the lectureship in Comparative Religion at Oxford and that he is still in Dachau. We have not been in any way approached to help get him out and the lady whose address Inge gave me did not reply to my letter ... We are still overwhelmed with claims of 500 Austrian scholars and scientists.[390]

It was clear that Eisler's links to the University of Oxford, through leading intellectuals such as Gilbert Murray, had enabled him to obtain a temporary appointment which was the prestigious Wilde Readership in Comparative and Natural Religion that had a tenure of three years. However, he was unable to take this up as he had been present in Austria during the Anschluss, had been arrested on 20 May 1938 and sent to the Dachau concentration camp. His wife Lili remained in Austria and worked to get him released. There were also major efforts made by Oxford University and the SPSL. On 21 February 1939, the Oxford Registrar Sir Douglas Veale had written to the SPSL:

> The University has taken considerable trouble to get Dr. Robert Eisler out of Germany. We have applied formally through the Embassy and

also tried various informal approaches, but there has been no response. I feel very doubtful if anything more can be done to help him.[390]

On 22 September 1938, Eisler had been transferred to the Buchenwald camp where he was made to do hard labour. He was, however, released in August 1939, miraculously just before the start of the war. He then managed to travel to Oxford via Italy and Switzerland.

On arriving in Oxford in September 1939, Eisler was disappointed to find that another scholar, Professor Edwin James, an anthropologist from the University of Leeds, had taken up the Wilde Readership in his absence. Eisler visited the vice-chancellor of the University and president of Magdalen College, Professor George Gordon, who told him that he thought Professor James would be prepared to give up the Wilde Readership as he had a permanent position back in Leeds University. However, James insisted on remaining in the post.[390] Gordon then became ill with cancer and died in March 1942. It seems that Eisler's appointment to the Wilde Readership was then forgotten.

With the war starting, many faculty members and students in Oxford joined the armed forces and numerous university and college buildings were requisitioned. The appeals to the vice-chancellor of a newly arrived refugee scholar from Austria could not have priority. However, Eisler was allowed to give some lectures and received a small stipend. He stayed initially with the ailing Sir Arthur Evans, the famous archaeologist who discovered the Palace of Knossos on Crete, before moving eventually to modest circumstances in Stow-on-the-Wold.[390]

Eisler's wife Lili had joined him in Oxford in the spring of 1940. However, like many other citizens of Greater Germany, on 25 June, he was interned after the German invasion of France and sent to the Isle of Man. Eisler had a friend, the diplomat and novelist Frank Ashton-Watkin, who worked hard to get him released. Eisler developed severe heart problems in the internment camp arising from his hard labour in Buchenwald and, as a result, he was released from internment on 5 September 1940.[390] His financial situation was now difficult and he wrote to the SPSL on 31 December 1940 requesting a grant to enable him to prepare a "general history of ancient, medieval and Renaissance astrology and astronomy." However, Tess Simpson's reply on 7 January 1941 was negative:

> I am afraid that it is not likely that the Allocation Committee will find it possible to give you a grant. It has never been the policy of the Society to give grants for research for its own sake; grants have been given to scholars whose prospects of early absorption were very good. Since the war has resulted in a severe decrease in our income, our Committee has had to be more rigid than ever on this point, and they are extremely reluctant to make new commitments. As a matter of fact, we are now compelled to cancel a number of existing grants.
>
> Mr. Adams was wondering whether it would not be possible for you to make some income by undertaking popular publications, e.g. in the Penguin-Pelican series? You have exceptional knowledge of subjects which we think would be of special interest to the editors of that series. Perhaps you have already made some efforts in this direction?[390]

Eisler, however, did not take the hint and wrote again to Tess Simpson on 11 January 1941 emphasising his extraordinary range of interests and influence:

> As to the Penguin-Pelican series I have a contract to write for them a popular book on Astrology. Not being a journalist, I have to build up a solid scientific foundation on the basis of which I can write the popular version. I have been hard at work all the last year upon the task ... But — especially owing to the loss of time by the more than two months internment — it will take me another three months hard work to get the manuscript ready for the publisher and I shall not get the £50 contracted until the proofs are corrected. I also offered a new popular edition of my new 850-page book on the Messiah Jesus but have not had an answer.
>
> Perhaps Mr. Adams could remind your Vice-President Sir William Beveridge that I have called upon him last autumn offering to give a course on the new (mathematical) theory of money and prices developed in two big French volumes by Dr. Georges Guillaume, Director of the Centre d' Études Économiques ... There is a great dearth of teachers and tutors here for economics. Nevertheless, nobody seems to remember

that the only foreign expert on monetary reform who has addressed — by invitation — both the British Parliamentary Finance Committee and the Committee on Banking and Currency of the U.S.A. Senate is here and most willing and eager to lecture on the subject.

I have seen the Master of Balliol too on the subject of the Nuffield College fellowships — founded to promote collaboration between practical business experience, such as I have accumulated in many years, and theoretical research, such as I have published in a series of books. But nothing has ever come of it. On the contrary, I have received more than one hint that by offering one's services one risks to create the impression of "pushing".[390]

During the war, it was very difficult for the Eislers to make ends meet and Lili even worked as a cook in return for the accommodation of two rooms.[391] This was somewhat degrading especially as they had once been well off financially, although Lili did appreciate the compliments on her cooking. Lili's valuable family property in Austria had been confiscated when Eisler was sent to Dachau and she refused to divorce him. Eisler had a younger brother Otto who was born in 1888. He had a career as a musician and a banker. However, in 1942, he was sent to the Maly Trostinets camp where he died.[392]

Eisler continued to write articles and gave occasional lectures on his wide variety of interests from ancient navigation to paper making. He published a major book entitled *The Royal Art of Astrology* which was a strong critique of astrology and astrologers.[393] Lindsay Anderson, who was then a student in Oxford and who was to make his name as a film director in the 1960s, went to hear Eisler speak on his book. On 22 January 1942, Anderson wrote in his diary on how he was charmed by the talk:

In the evening went to hear Dr. Robert Eisler speaking on "Astrology — Science or Superstition" at the Cosmos Society. He talked too much, in a delightful Austrian accent, but was erudite, so charming and had such an interesting subject but it didn't really matter.[394]

During his life, Eisler had often written letters to *The Times* and he continued this activity in the 1940s with letters on a variety of topics including Goethe's birthplace and on a letter he had acquired written by

Simon Bolivar in London.[395,396] Following the death of Hitler on 30 April 1945, Eisler wrote again to *The Times* to explain the origin of Hitler's name from the German word "Hütte" or "small cottager" which was "originally an expression of contempt on the part of the bigger landowners in the neighbourhood."[397] He stated at the end of his letter that he was "late of Dachau and Buchenwald."

On 20 March 1945, Ilse Ursell wrote to Eisler asking if he wished to apply for British citizenship. He replied somewhat brusquely that he did not want to change his allegiance to his home country of Austria. He also stated that he felt "utterly useless and unwelcome" in Oxford.[390] In April 1947, Ilse Ursell wrote her usual enquiry letter, this time to Sir Gilbert Murray, asking if he had information on the current whereabouts of Eisler. Murray had known Eisler for over 20 years but had become somewhat disenchanted by him. It was clear in Murray's reply, from his home Yatscombe, Boars Hill, Oxford, on 13 September 1947, that there were mixed views in Oxford about Eisler:

> Dr. Eisler is rather a problem. He is a man of immensely wide learning and a brilliant lecturer. Arthur Evans and I and some expert whose name I forget, warmly recommended him for the Wilde Lectureship. But he is rather slapdash and often wild in his theories and his more "methodisch" fellow countrymen are greatly down upon him and say he is a charlatan. It would be more true to say he is a very learned and somewhat reckless amateur.
>
> He is an Austrian Jew, and was formerly rich with, I believe, a beautiful estate. His wife is now a cook and he writes and does odd jobs. I am extremely sorry for him. One of his fellow prisoners on the Isle of Man said Eisler was a great nuisance, insisting on the best hour and room for lectures and so on, but that his lectures were about the most interesting he had ever heard.[390]

Eisler never held a permanent academic position in his whole career. His approach of regularly changing his interests from subject to subject did not fit in with the academic specialism expected at Oxford and, indeed, in the other great universities of Europe. His last major work was *Man into Wolf* in which he explored the idea that humans seek out not pleasures but strong sensations.[398] The footnotes took up many more pages than the text. The work was the subject of a lecture he gave to the Royal Society of Medicine. It is often credited as the first publication to use the term "serial killer," which was mentioned by Eisler in connection with Punch and Judy plays.[398,399]

Eisler's health had been poor since he had been in Dachau and Buchenwald with regular heart trouble. This most unusual of scholars died on 18 December 1949 at the age of 67 in an Arts and Crafts house, Chartfield, in Limpsfield in Surrey, where he and his wife had spent his last few years.[390] His obituary in *The Times* on 20 December, under the heading "An Encyclopaedic Author," was typically muted for that publication and described him as a "well-known writer on history and finance, though these subjects by no means exhausted the wide range of his interests."[400] A subsequent addition to the obituary from a family friend, Captain C. J. Goldsmid, was more effusive and stated on 18 January 1950, "As a lecturer in English he was supreme, developing abstruse argument with intriguing simplicity and clearness. Though possessed of encyclopaedic knowledge, he was entirely devoid of intellectual arrogance."[401]

In due course, his loyal wife Lili moved back to her family home at Unterach am Attersee in Upper Austria, after applying to the Austrian Government for its return. She lived there until her death on 17 March 1980 at the extraordinary age of 97.[402]

There are many similarities between Eisler and that other brilliant Austrian refugee who spent a period in Oxford, Erwin Schrödinger (1887–1961).[4] Both were born in Vienna in the 1880s and were brought up in quite wealthy families who had made their money from manufacturing. Schrödinger, who was five years younger than Eisler, was not Jewish but was considered a dissident by the German authorities after 1933. They went to top high schools in Vienna (Schrödinger to the Akademisches and Eisler to the Wasagasse) which left them with lifelong interest in the classics on which they both published. They enjoyed the liberal intellectual life of Vienna in the early 1900s when the arts, music and sciences were flourishing through great names such as Mahler, Klimt, Freud and Boltzmann. They both went to the University of Vienna where they were star pupils and obtained doctorates. They both served in the Austro-Hungarian army in World War I with some distinction. They were both associated with the leading intellectuals in Europe of their time — Albert Einstein, Ernst Cassirer, Karl Popper, Bertrand Russell, Gilbert Murray, Kenneth Clark and Sigmund Freud, to name just a few. They both lectured at the Eranos Institute — Eisler in 1935 on *The Fourth Gospel* and Schrödinger in 1946 on *The Spirit of Natural Science*.[403] However, there is no evidence, rather surprisingly, that Eisler and Schrödinger ever met or corresponded.

They both took up temporary positions abroad, with Schrödinger moving from Berlin to Magdalen College, Oxford, in 1933 and Eisler to Paris in 1925. With considerable naivety, they both returned to Austria not long before their country was taken over by the German forces in March 1938. After the Anschluss, both Eisler and Schrödinger were caught up with the Nazis (Eisler most severely) and both managed to escape via Italy and Switzerland to Oxford. There, they were treated with reserve by the conservative academic hierarchy. Neither managed to secure permanent positions in Oxford, with Schrödinger being promised the chair of a distinguished scholar in 1933 who failed to retire or die.[4]

At the end of 1938, Schrödinger went from Oxford to Ghent in Belgium. Then, with the outbreak of war in 1939, he found safety in Ireland.[4] Eisler stayed loosely connected with the University of Oxford from 1939 and spent the remaining period of his life somewhat unhappy. Schrödinger and Eisler were both assisted, through thick and thin, by very supportive wives. They both wrote on a bewildering range of subjects.

Schrödinger wrote one truly great paper which won him the Nobel Prize for Physics in 1933. His theory of wave mechanics nowadays forms the basis for a great deal of modern science, including the whole of chemistry, much of physics, materials science and molecular biology.[4] His name is known to every serious student of science. The brilliance of Eisler, however, only seems to have been rediscovered recently. In 2020, a sympathetic podcast series was produced that describes in detail the colourful life of this versatile but complicated polymath, and has the appropriate title that Eisler gave to himself, *A Very Square Peg*.[404]

Another exceptional but tragic and influential scholar who started in social sciences was the economist Paul Eppstein. He is the last scholar discussed in this book.

9.2 Paul Eppstein

Paul Eppstein (1902–1944) was an economist from Mannheim who became a Jewish Elder first in Berlin and then in the Theresienstadt ghetto. In 1944, he hosted a Red Cross visit that produced a positive report of living conditions in Theresienstadt, and has become the subject of much controversy.

Eppstein was born on 4 March 1902 in Ludwigshafen which is on the river Rhine in Germany. He studied a range of social sciences, including economics, law and political science at the University of Heidelberg. He obtained his doctorate there with his thesis having the intriguing title "The average as statistical diction (a contribution to statistical methodology based on the philosophy of the as-if)." In 1925, he became a lecturer in economics, law and political science at the Municipal College in Mannheim. There, he also set up a pioneering adult education centre and was the personal assistant to his doctoral supervisor Professor Salomon

Fig. 9.2   Paul Eppstein: *Reichsvertretung der Juden in Deutschland (1936–1938).*

Altmann at the University of Heidelberg.[405] In 1930, he married Hedwig Strauss.

Following the rise of the Nazis in 1933, Eppstein was dismissed from his positions in Mannheim. He then applied to the AAC in August 1933. All of his referees originated from Mannheim or Heidelberg and his application lacked the international aspect that many other applicants demonstrated. The AAC received a reference for Eppstein from Emil Lederer who had been Professor for Social Politics in Heidelberg and then Professor of the Humboldt University of Berlin. Lederer had escaped from Germany in 1933, first coming to London and then going on to the USA. There, he founded the New School for Social Research in New York which became a vital academic home for many refugees from Europe in social sciences.[376] On 25 February 1935, Lederer wrote to the AAC on Eppstein:

> I consider him a very well-trained man with excessive teaching experience. He was for years the assistant to Dr. Altmann at Mannheim and I know that he did a very good job there. Dr. Altmann was a very sick man for many years and Eppstein carried a good deal of his very great burden. It may be that he hasn't yet achieved any quite original work and his great erudition sometimes makes it difficult for him to take a decisive position but this caution was assured in his book *A Symptomatology of Business Cycles*, a very comprehensive study of the methodological problems evolved. I would rate him certainly "B" and I am pretty sure that he would have found under normal circumstances an adequate position in Germany.[406]

As we have seen for several other applicants, a "B" rating from an expert in the field was enough to ensure that the AAC would not give a priority to the application. Eppstein then took up a post in Berlin as the *Reichsvertretung der Juden in Deutschland* (Reich Representative of German Jews). This was an organisation which had been set up in September 1933 to coordinate the activities of Jewish groups in Germany including the provision of legal advice (see Figure 9.2). It was a prominent and dangerous role for Eppstein although the organisation was tolerated by the Nazis initially as it enabled a dialogue to be carried out with the

Jewish communities. In this role, he gave advice to many Jewish people in Berlin including Alfred Byk.[139] On behalf of the Jewish community, he also had to negotiate with Adolf Eichmann and other infamous Nazis.[407] In the latter half of the 1930s, and especially following the Kristallnacht of November 1938, Eppstein was put under much pressure and he was arrested several times by the Gestapo.[405]

Like several German Jewish academics in the social sciences, Eppstein was offered a position in the New School for Social Research in New York in 1939 but he delayed a move there as he wanted to complete his important work in Berlin.[405] In addition, the Dean of the School and previous referee for Eppstein, Emil Lederer, suddenly died in the same year. When the USA entered the war in 1941, this avenue was closed. On 27 January 1943, Eppstein and his wife were deported to Theresienstadt. There, he was appointed by Adolf Eichmann as Chair of the Jewish Council ("Aeltestenrat") due to his previous work with the Reich Representative in Berlin.[22,69] It is quite possible that he then had interactions with some of the other refugees discussed in this book who were deported to Theresienstadt during this period including Otto and Irma Sittig, Paul Funk, Hans Hirschfeld, Adalbert Kral, and Alexander and Jenny Duschinsky, the parents of Fritz Duschinsky.

On 23 June 1944, the Nazis arranged a unique Red Cross visit to Theresienstadt.[69,408,409] There is evidence that several thousand inmates of Theresienstadt had been deported to Auschwitz shortly before this visit to reduce the exceptional crowding in the camp.[408] As the Jewish Elder, Eppstein had the degrading task of escorting the visitors from Denmark and Switzerland along a carefully chosen route in the camp together with the closely watching SS officials (but not in uniform), including Karl Rahm, SS-Sturmbannführer and Commandant in Theresienstadt, and Dr. Erwin Weinmann, Commander-in-Chief of the German Security Police in Bohemia and Moravia. Eppstein was even provided with a car and a chauffeur for the visit and was given a prepared script on the "Jewish Self-Government in Theresienstadt."[410] The guests visited Danish Jews in their "apartments" which had been specially decorated and furnished for the visit. Eigil Henningsen was the Danish Red Cross member who wrote a summary of the visit:

Summarizing the numerous impressions that we received during our tour of Theresienstadt and supplementing these impressions with those of Dr. Eppstein and the German authorities, one comes to the opinion that great organizational progress has been made within this place closed off from the outside world, perhaps in the last six months, and that energy is still being worked on to improve where this is possible.

Economically, the population does not seem to be threatened by dangers, as long as there is granted a subsidy for the allocated rations through food consignments. The difficulties which the abnormal age distribution have presented to the administration also seem to be gone. One must have the deepest admiration for the Jewish leadership, and not least its Head Dr. Eppstein, who with blazing idealism has done everything to overcome all the difficulties of a material and spiritual kind.[411(t)]

Danish refugees were chosen as a focus for the Red Cross visit as King Christian X of Denmark had expressed concern about the fate of citizens in his country who had been transported to Theresienstadt. Another of the Red Cross representatives Maurice Rossel from Switzerland, who had recently spent time in Berlin, also wrote a report that was very positive about the camp and became even more controversial.[412] It included the following phrases:

We were told that Theresienstadt is the final repository for the delegates. In Geneva, I was repeatedly asked if inmates had been transferred from Theresienstadt to other ghettos in hundreds of cases. The Jews' elder Dr. Eppstein is the Stalin of Theresienstadt. He is the sole ruler along the Soviet method and rules very wisely thanks to his culture and cleverness. It is incomprehensible to the delegate that, for years, difficulties have been caused by the inspection of the camp, especially since the overall impression was so surprisingly good. The pictures I took with the permission of the camp commander were very important. Many interested circles would not have believed the report if the pictures (especially the children) could not have been presented.[413(t)]

Rossel also reported, "The overall impression of the settlement was very good. We were particularly impressed by the good appearance of the residents and the overall cleanliness."[413(t)] Rossel had previously made a controlled visit to Auschwitz on behalf of the Red Cross and, in a similar way to Theresienstadt, did not find anything that particularly perturbed him.[414]

The International Committee of the Red Cross was rightly praised for ensuring that prisoners of war were treated well and was awarded the Nobel Peace Prize in 1944.[415] However, it failed in communicating to the world the horrors of the concentration camps. The visit of the Red Cross Committee to Theresienstadt was part of this failure. After the war, the visit and the report of the Red Cross Committee were the subjects of critical articles, some of which also questioned Paul Eppstein's role.[408]

The Nazis thought that the Red Cross visit had gone so well that a film should be made about Theresienstadt for propaganda use. It was entitled "Theresienstadt: A Documentary Film from the Jewish Settlement Area" and featured Eppstein.[416,417] Some parts of the film are still preserved today. It has been suggested that one aim was to show the comfortable life of the people in Theresienstadt as compared to the hardships of the German citizens suffering regular allied bombing. The film showed women reading books and knitting while relaxing in the sunshine. It also portrayed elderly men in the camp playing chess. In addition, the film featured a football match in the courtyard of the camp surrounded by cheering spectators. The film was directed by a camp inmate and actor, Kurt Gerron, who had been transported from the Westerbork transit camp in the Netherlands, where the distinguished mathematicians Robert Remak and Otto Blumenthal had also been confined.[418] Shortly after the filming, Gerron and his wife Olga were deported on 28 October 1944 to Auschwitz on the same final transport as Professor Otto Sittig and his wife Irma. They were some of the last people to be murdered there in the gas chambers of Birkenau on 30 October 1944.[37,416]

Three months after the Red Cross visit, Paul Eppstein himself was arrested and shot in Theresienstadt on 27 September 1944.[405] There are several reasons that could explain why he was murdered, including a requirement to silence any critical comments he might make on the Red Cross visit, a possible refusal by him to help organise new transportations from Theresienstadt or the need to remove a trusted leader of the people in the ghetto.[405] On 28 October 1944, his wife Hedwig Eppstein was deported to Auschwitz on the very last transport, together with Otto and Irma Sittig and Kurt Gerron, where she also was murdered.[120,405] After the war, Karl Rahm, the Commandant of Theresienstadt who directed the Red Cross visit in 1944, was put on trial by the Czech government for war crimes and was found guilty. He was executed in 1947.[419]

On 20 September 1944, Tess Simpson wrote to Professor Arthur Salz, an economist in Columbus Ohio.[406] He had been a professor at Heidelberg and was assisted by the AAC in leaving Germany in 1933 and came to the University of Cambridge.[420] He then moved to the USA in 1934. Simpson asked if he knew anything about the whereabouts of Paul Eppstein who she hoped "had managed to get away from the Continent." Salz replied that Eppstein was "serving as a caretaker in the concentration camp in Theresienstadt."[406] Ilse Ursell, still unaware of Eppstein's fate, then wrote on 19 January 1948 to the Central Committee for Liberated Jews in the American Zone in Munich:

> According to information which we received in 1945, Dr. Eppstein was liberated at Theresienstadt where he had been serving with the "Aeltesten Rat". If Dr. Eppstein is still in Germany this Society may be able to submit his name for an opening in Palestine in his own field. We are anxious to have news of him as soon as possible.[406]

She then heard from Dr. E. Alexander of 41 Hertford Street in London on 21 January 1948: "I am told that a man of this name, who late in the 1930s was leading in the Reichsvertretung der Juden in Deutschland, has died at Theresienstadt."[406]

After the war, Eigil Henningsen, the chair of the Danish Red Cross committee which was shown around Theresienstadt by Eppstein, was accused of helping the Nazis cover up the Holocaust due to the non-critical report which he wrote on the visit.[411] Henningsen testified that his committee feared for the safety of Danish Jews if its report was too critical and he could only report what he had seen.[421] Maurice Rossel, the Swiss representative, was also interviewed several years later. He said he did not regret giving a positive impression of Theresienstadt in his report to the Red Cross which implied that the internees were not sent to death camps.[422]

Paul Eppstein's role as a Jewish Elder, first in Berlin and then in Theresienstadt, continues to be controversial.[412] At the trial of the infamous Adolf Eichmann in Jerusalem in 1961, following which Eichmann was executed by hanging, Paul Eppstein's name came up several times. This included discussions on his negotiations with Eichmann in Berlin on behalf of the Jewish community, his appointment to Theresienstadt as a Jewish Elder and his murder there in 1944.[423] In his evidence at the Eichmann trial in Jerusalem, Benno Cohn, who had been head of a pre-war Zionist organisation in Germany, gave a detailed account of a meeting between himself, Eppstein and Eichmann at the Gestapo Headquarters in Berlin late in 1938. Eichmann requested cooperation from Jewish agencies in setting up a Central Office for Jewish Emigration in Berlin based on one he had set up in Vienna. Cohn stated that Eppstein spoke very bravely: "You can of course throw us into a camp at any time, and do what you want with us, but as long as we are free you must recognise our dignity as human beings and treat us correspondingly."[407]

CHAPTER TEN

Conclusions

The Academic Assistance Council and its successor, the Society for the Protection of Science and Learning, played a crucial role in rescuing numerous scholars and scientists from Nazi persecution by facilitating their immigration to the UK. The importance of the AAC/SPSL cannot be underestimated. As the perceptive Rudolf Peierls stated, "In helping public opinion in avoiding misunderstanding and prejudice, the AAC and SPSL have given us help which was perhaps as important as the direct material assistance and advice which they have so generously provided."[424]

This book presents detailed accounts of thirty individuals and their families, whose fates or whereabouts were sought by the SPSL during or after World War II. Finding information on these refugees has been a major task, as many of them belonged to families that were wiped out by the Nazis, leaving no descendants. Furthermore, several of these individuals came from cities and institutes that were badly destroyed during the war, leaving minimal records behind. Hence, the well-preserved and meticulously maintained archives of the AAC/SPSL provided a vital source of comprehensive information on these scientists. Additionally, the police and immigration records from countries such as Belgium, Holland, France, Switzerland, and Czechoslovakia have also shed new light on the experiences of the refugees. By examining the stories of chemists, physicists, mathematicians, medical scientists, engineers, biologists and social scientists, this book uncovers common patterns that explain many of their tragic fates. The intriguing stories of Otto Frisch and Klaus Fuchs, who lost contact with the SPSL during the war due to their involvement in top secret work, are also briefly discussed.

The AAC and SPSL did not have substantial funds to assist refugees despite several fundraising appeals. The main effort was to find a suitable research fit for an applicant with a leading UK university or research institute. The links with the professors in those institutions were crucial as was the trust from the British Government in the integrity of the AAC/SPSL in only helping genuine scholars with research potential. As has been discussed in detail, the AAC/SPSL had to maintain a sensitive political position. The examples described here show that scientists in all the different subjects had to be known in the network of British professors and also among refugee scientists who had moved to the UK. They had to be of the right age. They had to be not too young like Duschinsky, whose brilliant work was not appreciated, or considered too old like Wreschner, Byk and several of the mathematical and medical scientists, as they would not be able to move on after a temporary visit to a UK institution had been arranged.

In Wreschner's case, her gender also lowered expectations as did her need to care for her mother. In the 1930s, science was very much a man's world. The subject of their research had to be of interest to the UK laboratories and the examples discussed here demonstrate that those whose area fell between the fields of chemistry and physics (including Duschinsky, Wasser, Wreschner, Byk, Pese and Lehmann, and the survivors in Belgium, Lasareff, Goldfinger and Rosen) had particular difficulty in being placed in the UK. Interdisciplinary fields of research have always presented special challenges and that was clearly the case here.

Some academic subjects had particular features which enabled special placings to be made. The growing chemical industries are an example where Paul Dreyfuss and others found employment in the area of colour photography. Furthermore, engineers were in demand outside of academia. In addition, some biology researchers were able to take up positions in distant countries with tropical climates which were not favoured by researchers in other fields. However, pure mathematicians did not have these advantages and unattractive teaching positions in schools were sometimes their only opportunity in the UK. Refugee medical scientists also had the British Medical Association opposed to their employment as doctors. The skills of some social scientists, such as lawyers and economists, were in demand although scholars in these areas did come up against the same challenges met by scientists in seeking positions in academia.

Another factor is that many temporary positions had been filled in the UK by 1938 so that few posts were available when distinguished academic refugees suddenly emerged from countries such as Austria and Czechoslovakia in the late 1930s. Several of the refugee scientists had spouses, children and, sometimes, parents to be cared for. In this book, some of these cases have been examined in detail and the family aspect made movements between countries more complicated.

The Soviet Union arises in several of the stories of the scientific refugees discussed in this book. A common factor amongst the scientists was the temptation to take up research positions there in the mid-1930s.[425] Duschinsky, Wasser, Houtermans and Simons all succumbed to this opportunity but with negative outcomes. They were offered the chance to direct their own independent research alongside outstanding colleagues in scientific centres such as Leningrad or Kharkov. It was a bitter disappointment to them that the politics in the USSR in the 1930s turned out to be as dangerous as in Germany. Tchernavin managed to escape from the Gulag in Russia and Riewe was taken after the war from Germany to work on the Soviet nuclear programme. Lustig, however, found a haven in Russia that served him well for the rest of his life.

Nearly all the applicants to the AAC/SPSL were Jewish but I found no evidence of anti-Semitism in the extensive AAC/SPSL set of documents which were examined in detail for this book. Nor was any such evidence found from the institutions or universities that assisted refugees wanting to come to the UK. This is not to say that there was no anti-Semitism among some other groups in the UK, with the Mosley black shirts being a prime example.[426] There was reserve from some British institutions such as the British Medical Association in receiving immigrants who may take jobs away from British doctors but this was not anti-Semitism. In the USA, however, there were documented examples of anti-Semitism in some universities.[13]

The fates or details of several of the scientists discussed here are incomplete in the AAC/SPSL files and those of other assistance agencies, in subsequent publications, on websites and sombre memorials. Accordingly, following extensive research, information has been provided in this book to bring their stories up to date. There are several findings that may cause surprise in some quarters including the details of the fates of Duschinsky and Wasser in France, the move of Lustig to Russia and Dreyfuss to Hollywood in California. Described also is the extraordinary and rarely told story of Riewe who, despite being a "non-Aryan," worked on the purification of uranium in both Nazi Germany and post-war Soviet Russia, and still survived to return to an important scientific administrative position in post-war West Germany. Some of the scientists discussed here, including Duschinsky, Lustig, Bemporad, Lasareff, Goldfinger, Rosen and Simons, bravely opposed the Nazis through war-related inventions, underground resistance groups or allied armies.

It is tragic that several of the individuals discussed in this book were caught up in the Nazi's killing machine of Jewish people right at its beginning and end. Alfred Lustig was transported from Vienna to Nisko in Poland in one of the very first deportations in October 1939. There, the arrangements were so chaotic that he managed to escape to a life in Russia where he became a war hero and an acclaimed university teacher. After this debacle in Poland, the transportations and death camps were then highly organised on an industrial scale for the next five years. The very last transportation to Auschwitz on 28 October 1944 included five people discussed in this book: the distinguished medical scientist Otto Sittig and his wife Irma, Hedwig, the wife of Paul Eppstein, and the film director Kurt Gerron and his wife Olga. Adolf Eichmann was central in the transportation arrangements, and the social scientist Paul Eppstein, the last of the scholars considered in this book, had to negotiate directly with Eichmann in Berlin and Theresienstadt as a Jewish Elder.

Even in 2022 and 2023, new memorials or street namings to remember people who were murdered in World War II continue to be established. Several examples are given in this book of such recent commemorations of distinguished scientists. It appears that some countries are only now beginning to fully comprehend the atrocities committed in the death camps, some 80 years after the horrific events.

In order to escape from countries under the rule of the Hitler regime, undertaking academic research served as a means of finding assistance. Within this book, we have thoroughly investigated the limited number of scientists who were unable to obtain aid from UK assistance organisations, despite the successful help extended to hundreds of other scholars to relocate to the UK.[2,3,6,8,9] There were also numerous scientists who did not attempt to come to the UK and were murdered by the Nazis.[13,107,108,109,280,292,296,299] In addition, there were, of course, millions of people who did not have the academic escape route and perished in the Holocaust.[67]

Right to the present day, the ability of countries to deal with refugees remains a major challenge. The AAC was established in 1933 and was reformed as the SPSL in 1936. The organisation has been adapted several times since then as times and needs have changed. The modern title is the Council for At-Risk Academics (CARA) which, in the current century, has been assisting academic refugees from several countries ravaged by war including Iraq, Afghanistan, Syria and the Ukraine. Some of these recent refugees are established scientists and the hurdles they have been facing are similar to those of the Lost Scientists of World War II. It is hoped that appreciation of the challenges faced by the individuals discussed in this book will be of value in assisting refugees in the present day and in the future.

References

MS. S.P.S.L. refers to the Archive of the Society for the Protection of Science and Learning, Bodleian Library, Oxford.

Chapter 1: Crisis in Europe in the 1930s

1. N. T. Greenspan, *The End of the Certain World: The Life and Science of Max Born* (Basic Books, New York, 2005).
2. W. Beveridge, *A Defence of Free Learning* (Oxford University Press, 1959).
3. R. M. Cooper (Ed.), *Refugee Scholars. Conversations with Tess Simpson* (Moorland Books, Leeds, 1992).
4. D. C. Clary, *Schrödinger in Oxford* (World Scientific Publishing, Singapore, 2022).
5. Archive of the Society for the Protection of Science and Learning, Bodleian Library, Oxford.
6. D. Zimmerman, The Society for the Protection of Science and Learning and the politicization of British science in the 1930s. *Minerva*, **44**, 25 (2006).
7. J. Seabrook, *The Refuge and the Fortress: Britain and the Flight from Tyranny* (Palgrave Macmillan, London, 2008).
8. N. Bentwich, *The Rescue and Achievement of Refugee Scholars: The Story of Displaced Scholars and Scientists 1933–1952* (M. Nijhoff, The Hague, 1953).
9. J. Medawar and D. Pyke, *Hitler's Gift: The True Story of the Scientists Expelled by the Nazi Regime* (Arcade, New York, 2000).
10. J. Borkin, *The Crime and Punishment of I.G. Farben* (Barnes and Noble, New York, 1978).
11. A. Roberts, *The Storm of War: A New History of the Second World War* (Allen Lane, London, 2009).

12. Emergency Committee in Aid of Displaced Foreign Scholars Records, 1927–1949, Manuscripts and Archives Division, The New York Public Library.

13. L. Leff, *Well Worth Saving. American Universities' Life-and-Death Decisions on Refugees from Nazi Europe* (Yale University Press, New Haven, 2019).

14. S. Lee and G. E. Brown, Hans Albrecht Bethe 2 July 1906–6 March 2005, *Biogr. Mems Fell. R. Soc.*, **53**, 1 (2007).

15. A. Bhattacharya, *The Man From the Future: The Visionary Life of John von Neumann* (Allen Lane, London, 2021).

16. K. Gavroglu, *Fritz London: A Scientific Biography* (Cambridge University Press, 1995).

17. E. P. Wigner, *The Recollections of Eugene P. Wigner, as told to Andrew Szanton* (Springer, New York, 1992).

18. G. Kreft, 'Dedicated to represent the true spirit of the German nation in the world': Phillipp Schwartz (1894–1977), founder of the Notgemeinschaft, Chapter 8, in S. Marks, P. Weindling and L. Wintour (Eds.) *In Defence of Learning: The Plight, Persecution and Placement of Academic Refugees, 1933–1980s* (Oxford University Press for the British Academy, 2011).

19. C. Dyhouse, The British Federation of University Women and the status of women in universities, 1907–1939. *Women's Hist. Rev.*, **4**, 465 (1995).

20. S. Cohen, Defence of academic women refugees: The British Federation of University Women, Chapter 10, in S. Marks, P. Weindling and L. Wintour, (Eds.) *In Defence of Learning: The Plight, Persecution and Placement of Academic Refugees, 1933–1980s* (Oxford University Press for the British Academy, 2011).

Chapter 2: Physics and Chemistry Non-Survivors

21. U. Deichmann, The expulsion of Jewish chemists and biochemists from academia in Nazi Germany. *Perspect. Sci.*, **7**, 1 (1999).

22. A. Dagan, G. Hirschler and L. Weiner (Eds.) *The Jews of Czechoslovakia. Historical Studies and Surveys*, Vol. III (Jewish Publication Society of America and the Society for the History of Czechoslovak Jews, Philadelphia, 1984).

23. Erich Duschinsky interviewed by Mark Burman, Central British Fund Kindertransport Interviews, British Library Recordings, 1988.

24. Duschinsky, Dr. Franz (1907-), File 1934-, MS. S.P.S.L. 480/3.

25. R. Singh and F. Reiss, Seventy years ago — The discovery of the Raman effect as seen from German physicists. *Curr. Sci.*, **74**, 1112 (1998).

26. F. Duschinsky, Der Einfluss von Zusammenstössen auf die Abklingzeit der Na-Resonanzstrahlung (On the influence of collisions on the lifetime of excited Na atoms). *Z. Phys.*, **78**, 586 (1932).

27. F. Duschinsky, Der Zeitliche Intensitätsverlauf von Intermittierend Angeregter Resonanzstrahlung (The temporal course of the intensity of intermittently excited resonance radiation). *Z. Phys.*, **81**, 7 (1933).

28. F. Duschinsky, Eine Allgemeine Theorie der Zur Messung Sehr Kurzer Leuchtdauern Dienenden Versuchsanordnungen (Fluorometer) (A general theory of experimental arrangements (fluorometers) for measuring very short light durations). *Z. Phys.*, **81**, 23 (1933).

29. J. R. Lakowicz, *Principles of Fluorescence Spectroscopy* (Springer Verlag, New York, 2006).

30. Duschinsky to Schrödinger, 11 November 1933, Schrödinger Archiv, Österreichische Zentralbibliothek für Physik, Wien.

31. Peter Pringsheim, Foreigners' File No. 1.146.765, National Archives of Belgium.

32. M. Farr, *Tintin & Co.* (Egmont, UK, 2007).

33. Duschinsky, Fritz (1933–1940), Box 53.11, Emergency Committee in Aid of Displaced Foreign Scholars Records, New York Public Library.

34. Friedrich Duschinsky, Foreigners' File No. A106.433, National Archives of Belgium.

35. F. Duschinsky and P. Pringsheim, Ultraviolette Fluoreszenzspektra des Joddampfes: Einfluss von Fremdgasen (Ultraviolet fluorescence spectra of iodine vapour: Influence of foreign gases). *Physica*, **2**, 923 (1935).

36. W. E. Garner and J. E. Lennard-Jones, Molecular spectra and molecular structure. A general discussion. *Trans. Faraday Soc.*, **25**, 611 (1929).

37. Yad Vashem Archives, Jerusalem, Israel. Personal Documents: Duschinsky Fritz, ID 14151354-5.

38. A. Kojevnikov, President of Stalin's Academy: The mask and responsibility of Sergei Vavilov. *Isis*, **87**, 18 (1996).

39. F. Duschinsky, Über Prädissoziation bei Mehratomigen Molekülen (The predissociation in polyatomic molecules). *Acta Physicochimica URSS*, **5**, 651 (1936).

40. F. Duschinsky, "Abnormal" anti-Stokes fluorescence. *Comptes Rendus de L'Académie des Sciences de l'URSS*, **17**, 179 (1937).

41. F. Duschinsky, Zur Deutung der Elektronenspektren mehratomiger Moleküle. I. Uber das Franck-Condon-Prinzip (The interpretation of the

electronic spectrum of polyatomic molecules. I. On the Franck-Condon principle). *Acta Physicochimica URSS*, **7**, 551 (1937).

42. M. Volkenstein, M. Eliashevich and B. Stepanov, The theory of vibrational spectra of polyatomic molecules. 1. The problems of theory. *Z. Eksp. Teoret. Fiz.*, **15**, 35 (1945).

43. H. Altmann, *Improvements in Reflectors*, 9 January 1936 (G.B. Patent 450,076).

44. M. R. Lazewski, Virtuelle Rekonstruktion der Synagoge Jablonec nad Nisou (Gablonz an der Neisse) (Virtual reconstruction of the synagogue Jablonec nad Nisou (Gablonz an der Neisse)), Thesis, Technische Universität Wien, publik.tuwien.ac.at/files/PubDat_227736.pdf.

45. M. Altmann and F. Duschinsky, *Appareillage de Navigation (Navigation Equipment)*, 1 April 1939 (French Patent 857,562).

46. H. Altmann, M. Altmann and F. Duschinsky, *Systèmes Autocollimateurs (Autocollimater Systems)*, 22 August 1939 (French Patent 862,278).

47. S. Hermann, M. Altmann and F. Duschinsky, *Dispositif Augmentant l'Efficacité des Masques à Gaz (Device Increasing the Efficiency of Gas Masks)*, 16 October 1939 (French Patent 863,851).

48. M. Altmann, F. Duschinsky and S. Hermann, *Dispositif Pour Diriger de la Lumière de Luminescence dans la Direction de la Source de la Lumière Excitatrice (Device for Directing Luminescent Light in the Direction of the Exciting Light Source)*, 18 October 1939 (French Patent 863,859).

49. Siegwart Hermann Obituary. *New York Times*, 28 November 1956.

50. V. Mastny, The Czechoslovak government-in-exile during World War II. *Jahrbücher für Geschichte Osteuropas*, **27**, 548 (1979).

51. S. Hermann, M. Altmann and F. Duschinsky, *Signalisation par Feux et Fumées (Fire and Smoke Signalling)*, 1 November 1939 (French Patent 863,925).

52. S. Hermann, M. Altmann and F. Duschinsky, *Dispositif d'Éclairage (Lighting Device)*, 21 November 1939 (French Patent 864,204).

53. Société d'Exploitation des Verreries de Bagneaux et Appert Frères Réunie and H. Altmann, *Appareil Réflecteur Particulièrement Destiné à la Signalisation (Reflector Device Particularly Intended for Signalling)*, 1 February 1941 (French Patent 873,098).

54. Société d'Exploitation des Verreries de Bagneaux et Appert Frères Réunie, *Procédé et Appareil pour la Surfaçage de Corps Sphériques, en Particulier de Verres d'Optique (Method and Apparatus for Surfacing Spherical Bodies, in Particular Optical Glasses)*, 21 October 1942 (French Patent 977,720).

55. Cornings, www.corning.com/worldwide/en/the-progress-report/crystal-clear/cornings-18th-century-plant-forms-the-newest-glass-technologies.

56. S. Hermann, *Verfahren zur Herstellung von Therapeutisch Wirksamen Präparation mit Hilfe von Kombucha (Process for the Manufacture of a Therapeutically Effective Preparation Using Kombucha)*, 20 February 1927 (German Patent 538,028).

57. R. Kolkwitz, *Pflanzenphysiologie: Versuche und Beobachtungen an Höheren und Niederen Pflanzen Einschliesslich Bakteriologie und Hydrobiologie mit Planktonkunde (Plant Physiology: Experiments and Observations on Higher and Lower Plants including Bacteriology and Hydrobiology with Plankton Science)* (Gustav Fischer, Jena, 1914).

58. The Statue of Liberty-Ellis Island Foundation, Passenger Lists of Vessels Arriving at New York.

59. S. Hermann, *System and Device for Signaling Purposes*, 10 May 1941 (US Patent 2,334,765).

60. S. Hermann, *Gas Mask*, 23 May 1941 and in France 16 October 1939 (US Patent 2,284,053).

61. S. Klarsfeld, *Le Calendrier de la Persécution des Juifs de France. Septembre 1942–Août 1944. La Shoah En France (The Calendar of the Persecution of the Jews of France. September 1942–August 1944. The Shoah in France)* (Fayard, Paris, 2001).

62. To the deportees of Bagneaux-sur-Loing, 8 May 2022, www.cercleshoah.org/spip.php?article1034.

63. United States Holocaust Memorial Museum, Data compiled by Georg Dreyfuss regarding deportees from France, based on Serge Klarsfeld, *Le Mémorial de la Déportation des Juifs de France (The Memorial to the Deportation of Jews from France)*, ID: 20564.

64. Convoi 45 parti du camp de Drancy le 11/11/1942 à destination du camp d'Auschwitz-Birkenau (Convoy 45 from Drancy to Auschwitz-Birkenau 11/11/1942), C45_21: p. 21, *Mémorial de la Shoah, Paris (France)*.

65. S. Klarsfeld, *Die Endlösung der Judenfrage in Frankreich. Deutsche Dokumente 1941–1944 (The Final Solution of the Jewish Question in France. German Documents 1941–1944)* (Klarsfeld, Paris, 1977).

66. S. Klarsfeld, *Le Mémorial de la Déportation des Juifs de France, Avis de Convois (The Memorial to the Deportation of Jews from France, Notice of Convoys)* (1978). See also www.lesdeportesdesarthe.wordpress.com/convoi-n-45-du-11-novembre-1942/.

67. Yad Vashem Archives, Jerusalem, Israel. Central Database of Shoah Victims' names.

68. Sterbebücher von Auschwitz (Death Book from Auschwitz), *Staatliches Museum Auschwitz-Birkenau* (De Gruyter Saur, Munich, 1995).

69. H. G. Adler, *Theresienstadt 1941–45. Das Antlitz einer Zwangsgemeinschaft, Geschichte Soziologie Psychologie (Theresienstadt 1941–44. The Face of a Compulsory Community, History Sociology Psychology)* (Mohr, Tübingen, 1955).

70. S. L. Wolff, Entrechtet, Verfolgt, Vertrieben und Ermordet. Auch Mitglieder der DPG Wurden Opfer des Nationalsozialistischen Systems (Disenfranchised, Persecuted, Expelled and Murdered Members of the DPG Also Became Victims of the National Socialist System), *Physik Journal*, **19**, 29 (2020).

71. M. Yamaguchi, Life of Fritz Duschinsky and photochemistry. *J. Jap. Soc. Hist. Chem.*, **45**, 117 (2018).

72. Q. Peng, Y. Yi, Z. Shuai and J. Shao, Excited state radiationless decay process with Duschinsky rotation effect: Formalism and implementation. *J. Chem. Phys.*, **126**, 114302 (2007).

73. J. B. Coon, R. E. Dewames and C. M. Loyd, The Franck-Condon principle and structures of excited electronic states of molecules. *J. Mol. Spectrosc.*, **8**, 285 (1962).

74. G. Tian, S. Duan, G.-P. Zhang, W. Hu and Y. Luo, The effect of Duschinsky rotation on charge transport properties of molecular junctions in the sequential tunneling regime. *Phys. Chem. Chem. Phys.*, **17**, 23007 (2015).

75. R. Ianconescu and E. Pollak, Photoinduced cooling of polyatomic molecules in an electronically excited state in the presence of Dushinskii rotations. *J. Phys. Chem. A*, **108**, 7778 (2004).

76. F. Ehrenhaft and E. Wasser, Grössen-, Gewichts- und Ladungsbestimmung Submikroskopischer Einzelner Kugeln der Radiengrössen: 4.10^{-5} cm bis 5.10^{-6} cm, Mit Reeller Abbildung Submikroskopischer Teilchen durch das Ultraviolette Licht (Determining the size, weight and charge of submicroscopic single spheres with radii sizes 4×10^{-5} to 5×10^{-6} cm, with real imaging of submicroscopic particles by ultraviolet light). *Z. Phys.*, **37**, 820 (1926).

77. F. Ehrenhaft and E. Wasser, Determination of the size and weight of single submicroscopic spheres of the order of magnitude $r = 4.10^{-5}$ cm to

5.10^{-6} cm, as well as the production of real images of submicroscopic particles by means of ultraviolet light. *Phil. Mag.*, **7**(2), 30 (1926).

78. G. Magalhães Santos, A debate on magnetic current: The troubled Einstein-Ehrenhaft correspondence. *Brit. J. Hist. Soc.*, **44**, 371 (2011).

79. Ehrenhaft to Einstein 6 December 1926, *The Collected Papers of Albert Einstein. The Berlin Years: Writings and Correspondence June 1925–May 1927*, Vol. 15, p. 943 (Princeton University Press, 2018).

80. O. Halpern and E. Wasser, A direct experimental test of the principle of spectroscopic stability. *Phys. Rev.*, **46**, 177 (1934).

81. Wasser, Dr. Emmanuel (1903-), File 1935-48, MS. S.P.S.L. 342/6.

82. Fröhlich, Dr. Herbert (1905–91), File 1933-48, MS. S.P.S.L. 328/1.

83. N. Mott, Herbert Fröhlich 9 December 1905–23 January 1991, *Biogr. Mems Fell. R. Soc.*, **38**, 147 (1992).

84. Ribbentrop to Henderson, 29 April 1938, *Home Office. Schrödinger, Erwin Rudolf and wife Anna Maria: Erwin Schrödinger, an Austrian physicist*. HO 382/105, UK National Archives Kew.

85. Ehrenhaft, Prof. Felix (1879-), File 1938-44, MS. S.P.S.L. 326/4.

86. D. C. Clary, Foreign Membership of the Royal Society: Schrödinger and Heisenberg? *Notes Rec.*, **77**, 513 (2023).

87. Reiss, Dr. Max (1903-), File 1938-48, MS. S.P.S.L. 337/7.

88. Auguste Piccard, Explorer, is dead. *New York Times*, 26 March 1962.

89. Emmanuel Wasser, Foreigners' File No. A326.560, National Archives of Belgium.

90. P. Pringsheim, Fluorescence and phosphorescence of thallium-activated potassium halide phosphors. *Rev. Mod. Phys.*, **14**, 132 (1942).

91. M. Bervoets, *La Liste de St. Cyprien (The List of St. Cyprien)* (Alice Editions, Brussels, 2006).

92. Emanuel Wasser, *U.S., Jewish Transmigration Bureau Deposit Cards, 1939–1954 (JDC)*, Provo, UT, USA. Roll 04 — Case 7853.

93. Y. Bauer, *American Jewry and the Holocaust: The American Jewish Joint Distribution Committee, 1939–45* (Wayne State University Press, 1981).

94. M. Arnold, La Rafle du 25 Novembre 1943 (The Round-Up of 25 November 1943). *Rev. Hist. Phil. Relig.*, **3**, 353 (2011).

95. De l'Université aux Camps de Concentration: Témoignages Strasbourgeois (From University to Concentration Camps: Testimonies from Strasbourg) (Presses Universitaires de Strasbourg, 1995).

96. Państwowe Muzeum w Oświęcimiu (Staatliches Museum in Auschwitz), Deportation list Nr. 66: Transport from 20.01.1944 from Drancy to Auschwitz. Document number 11183860, Arolsen Archives.

97. Dossier: Le Convoi du 20 Janvier 1944. *Mémoire Vivant*, **65**, 1 (2010).

98. Journal Officiel, République Française, Lois et Décrets (Official Journal, French Republic, Laws and Decrees), Order of January 28, 2002 affixing the words "Death in deportation" to death certificates and judgments, Text No. 45, 24 March 2002.

99. S. Birnbaum, *Une Française Juive est Revenue: Auschwitz, Belsen, Raguhn (A Jewish Frenchwoman Returns: Auschwitz, Belsen, Raguhn)* (Hérault, Maulévrier, 1989).

100. Wasser, Elisabeth, Police File N21992, Swiss Federal Archives, Federal Department of Home Affairs, Berne.

101. C. Hoff, *Anna und Leon* (Hentrich & Hentrich, Berlin, 2005).

102. French Republic, File of Deceased Persons, www.data.gouv.fr/fr/datasets/fichier-des-personnes-decedees/.

103. Dreyfuss, Dr. Paul (1906-), File 1934-, MS. S.P.S.L. 480/1-4.

104. Universitat de Strasbourg Plaque Commémorative, museedelaresistance enligne.org/media2860-Universit-de-Strasbourg-plaque-commmorative.

105. P. A. M. Dirac, Ehrenhaft, the Subelectron and the Quark, in C. Weiner (Ed.) *History of Twentieth Century Physics: Proceedings of the International School of Physics Enrico Fermi. Course LVII (1972)*, p. 290 (Academic Press, New York, 1977).

106. P. A. M. Dirac, The theory of magnetic poles. *Phys. Rev.*, **74**, 817 (1948).

107. R. Rurup and M. Schüring, *Schicksale und Karrieren. Gedenkbuch für die von den Nationalsozialisten aus de Kaiser-Wilhelm-Gesellschaft vertriebenen Forscherinnen und Forscher.* Vol. 14 (*Destinies and Careers. Memorial book for the Researchers Expelled from the Kaiser Wilhelm Society by the National Socialists. Series: History of the Kaiser Wilhelm Society under National Socialism*, Vol. 14) (Wallstein Verlag, Göttingen, 2008).

108. J. James, T. Steinhauser, D. Hoffmann and B. Friedrich, *One Hundred Years at the Intersection of Chemistry and Physics. The Fritz Haber Institute of the Max Planck Society 1911–2011* (De Gruyter, Berlin, 2011).

109. A. Vogt, *Wissenschaftlerinnen in Kaiser-Wilhelm-Instituten, A–Z (Female Scientists in Kaiser Wilhelm Institutes, A–Z)*, pp. 218–220 (Archiv der Max-Planck-Gesellschaft, Berlin, 2008).

110. H. Freundlich and M. Wreschner, Ueber den Einfluss der Farbstoffe auf die Elektrokapillarkurve (On the influence of dyes on the electrocapillary curve). *Kolloid Z.*, **28**, 250 (1921).

111. M. Wreschner and L. F. Loeb, *Composition of Matter and Method of Producing Same*, 1 April 1930 (USA Patent 1,752,826).

112. R. Robinson, *Versuch Einer Elektronentheorie Organisch-Chemischer Reaktionen (Attempt at an Electron Theory of Organic Chemical Reactions)*, transcription from the English by Dr. M. Wreschner (Verlag Enke, Stuttgart, 1932).

113. Söllner, Dr. Karl (1901-), File 1922-48, MS. S.P.S.L. 225/2.

114. E. Abderhalden, *Handbuch der biologischen Arbeitsmethoden (Handbook of Biological Working Methods)* (Berlin und Wien, Urban und Schwarzenberg, 1929).

115. Wreschner, Dr. Marie (1887-), File 1938-47, MS. S.P.S.L. 343/5.

116. Foreign Members Book, MS/719, Royal Society Library, London.

117. Records of the British Federation of University Women, London University: London School of Economics, the Women's Library, Ref. No. 5BFW/12/26 Box 159, Wreschner, Dr. Marie, 1939.

118. R. Huang, *A Lifetime in Academia: An Autobiography of Rayson Huang* (Hong Kong University Press, 2011).

119. Max Bergmann papers, Mss. B. B445, Box 23, Folder 24, American Philosophical Society Library.

120. A. Gottwaldt and D. Schulle, *Die 'Judendeportationen' aus dem Deutschen Reich von 1941–1945 (The 'Jewish Deportations' from the German Reich 1941–1945)* (Marix Verlag, Wiesbaden, 2005).

121. R. P. Miller, Auflistung und Einführung zur Liste der am 14. November 1941 aus Berlin in das Ghetto Minsk Deportierten Jüdinnen und Juden, Aus Berlin-Minsk: Unvergessene Lebensgeschichten *(Introduction to the List of Jews Deported from Berlin to the Minsk Ghetto on November 14, 1941, From Berlin-Minsk: Unforgotten Life Stories)*, A. Reuss and K. Schneider (Eds.) (Metropol Verlag, Berlin, 2013).

122. Marie Wreschner (20/9/87), Zählkarte für Zuzug, Fortzug, Sterbefall (Counting Card for Arrival, Departure, Death), 17/11/1941, Doc. ID. 12679367, ITS Digital Archive, Arolsen Archives.

123. Marie Sara Wreschner, Sterberegister der Berliner Standesämter 1874–1955 (Death Register of the Berlin Registry Offices 1874–1955), Landesarchiv, Berlin, Deutschland.

124. M. Born, Victims of the Nazis Amongst my Relatives and Friends, GBR/0014/BORN/1/2/1/10, Papers of Professor Max Born, Churchill Archives Centre, Churchill College, Cambridge.

125. Bezirksverordnetenversammlung Marzahn-Hellersdorf von Berlin, Frauennamen für Strassenbenennungen, hier: Bereich des Clean Tech

Business Parks Ausschussantrag (District assembly Marzahn-Hellersdorf from Berlin, Women's Names for Street Names: Area of the Clean Tech Business Park Committee Proposal), 1362/VII-1, 26 February 2015.

126. The women who made Pacific: Cecilie Fröhlich. *Pacific Magazine,* Pacific University Oregon, Fall, 2019. Pacific.edu/magazine.

127. A. B. Vogt, Gertrud Kornfeld (1891–1955), in J. Apotheker, and L. S. Sarkadi (Eds.) *European Women in Chemistry* (Wiley, Weinheim, 2011).

128. Kornfield, Dr. Gertrud (1891–1955), File 1933-38, MS. S.P.S.L. 218/3.

129. G. Kornfeld and M. McCaig, The photochemical decomposition of sulphur dioxide. *Trans. Faraday Soc.,* **30**, 991 (1934).

130. G. Kornfeld, Some new ultraviolet bands of SO_2 in emission. *Trans. Faraday Soc.,* **32**, 1487 (1936).

131. G. Kornfield, Limits of infra-red sensitizing. *J. Chem. Phys.,* **6**, 201 (1938).

132. E. Serrano, J. Mercelis and A. Lykknes, 'I am not a lady, I am a scientist.' Chemistry, women, and gender in the enlightenment and the era of professional science. *Ambix, J. Soc. Hist. Alch. Chem.,* **69**, 203 (2022).

133. A. Byk, Uber Einige Derivate des Pyrimidins (On some derivatives of pyrimidine). *Ber. Der Deutschen. Chem. Gesell.,* **36**, 1915 (1903).

134. A. Byk, Zur Frage der Spaltbarkeit von Razemverbindungen Durch Zirkular Polarisiertes Licht, ein Beitrag zur Primären Entstehung Optisch-Aktiver Substanz (The fissionability of racemic compounds by circularly polarised light, on the primary appearance of an optically active substance). *Z. Phy. Chem.,* **49**, 641 (1904).

135. World War I Document Archive, Brigham Young University, USA. See also www.gwpda.org/1914/profeng.html.

136. Private communication to the author from Professor R. Pattenden.

137. M. Born and A. Einstein, *The Born-Einstein Letters 1916–1955* (Macmillan Press, New York, 1971).

138. W. Heitler and F. London, Wechselwirkung Neutraler Atome und Homöopolare Bindung Nach der Quantenmechanik (Interaction of neutral atoms and homopolar bonds according to quantum mechanics). *Z. Phys.,* **44**, 455 (1927).

139. Byk, Alfred (1933–1934, 1939), Box 48.32, Emergency Committee in Aid of Displaced Foreign Scholars Records, New York Public Library.

140. Byk, Professor Alfred (1878–c1945), File 1934-48, MS. S.P.S.L. 475/1.

141. Private collection of Professor R. Pattenden.

142. Anna Ernestine Graeffner, Sterberegister der Berliner Standesämter 1874–1955 (Death Register of the Berlin Registry Offices 1874–1955), Landesarchiv, Berlin, Deutschland.

143. D. Templin, Auftrag des Staatsarchivs Hamburg Wissenschaftliche Untersuchung zur NS-Belastung von Strassennamen Abschlussbericht (Commissioned by the Hamburg State Archives, Scientific Investigation into the Nazi Burden of Street Names, Final Report), 2017.

144. Hedwig Amalie Sara Byk, Sterberegister der Berliner Standesämter 1874–1955 (Death Register of the Berlin Registry Offices 1874–1955), Landesarchiv, Berlin, Deutschland.

145. S.L. Wolff, Alfred Byk (1878–1942). *Phys. J.*, **19**, 35 (2020).

146. Behrend, Dr. Felix Adalbert (1911–1962), File 1933-47, MS. S.P.S.L. 277/4.

147. B. H. Neumann, Felix Adalbert Behrend. *J. Lond. Math. Soc.*, **S1-38**, 308 (1963).

148. Berlin Ehrt Ermordeten Wissenschaftler: Gedenktafel für Alfred Byk (Berlin honours murdered scientist: Memorial plaque for Alfred Byk). *Berliner Zeitung*, 31 January 2023.

149. S. L. Wolff, Herbert Pese (1899–1943). *Phys. J.*, **19**, 53 (2020).

150. C. Schaefer and H. Pese, Dur Definition von Sättigung (The definition of saturation). *Physik. Z.*, **31**, 720 (1930).

151. Pese, Dr. Herbert (1899-), File 1938-48, MS. S.P.S.L. 336/3.

152. A. Ascher, *A Community under Siege. The Jews of Breslau under Nazism* (Stanford University Press, 2007).

153. R.B. Müller, Vom Ende des Jüdischen Schulwesens in Breslau (The end of the Jewish school system in Breslau). *Medaon (Magazine for Research in Jewish Life and Education)*, **1** (2007).

154. Pese, Herbert, Gedenkbuch Opfer der Verfolgung der Juden unter der Nationalsozialistischen Gewaltherrschaft in Deutschland 1933–1945 (Memorial Book of Victims of the Persecution of the Jews under National Socialist Tyranny in Germany 1933–1945), Bundesarchiv, Koblenz.

155. S. L. Wolff, Erich Lehmann (1878–1942). *Phys. J.*, **21**, 22 (2022).

156. E. Ladenburg and E. Lehmann, Über Versuche mit Hochprozentigem Ozon (Experiments with high percentage of ozone). *Ann. der Phys.*, **21**, 305 (1906).

157. A. Miethe and E. Lehmann, Über das Ultraviolette Ende des Sonnenspektrums, (Concerning the ultraviolet end of the Sun's spectrum). *Sitz. Der Königlich Preuss. Ak. Der Wiss.*, **7**, 268 (1909).

158. E. Lehmann, Über das Verhältnis von Absorption und Empfindlichkeit bei Photographischen Präparaten (The proportions of absorption and sensitivity in photographic compounds). *Z. Phys. Chem.*, **64**, 89 (1908).

159. J. Eggert, in *Photographische Korrespondenz, 6. Sonderheft: Berlin und seine Bedeutung für die Photochemie in Wissenschaft und Technik für die Photoindustrie und für die Photowirtschaft* (*Photographic Correspondence, 6th Special Issue: Berlin and its Importance for Photochemistry in Science and Technology for the Photo Industry*), p. 27, (1964).

160. C. Rath and E. Lehmann, Über Cis- und Trans- Isomerie in der Reihe der Stilbazole (Concerning cis and trans isomers in the stilbazol series). *Ber. der Deutsch. Chem. Gesell.*, **58**, 342 (1925).

161. Lehmann, Professor Erich (1878–c1940), File 1939-43, MS. S.P.S.L. 516/1.

162. Victor Lehmann, in *Association of Jewish Refugees Information*, **XVI**, 13 (1961).

163. Lehmann, Erich (1934, 1941), Box 87.9, Emergency Committee in Aid of Displaced Foreign Scholars Records, New York Public Library.

164. Margarethe Sara Jacoby, Sterberegister der Berliner Standesämter 1874–1955 (Death Register of the Berlin Registry Offices 1874–1955), Landesarchiv, Berlin, Deutschland.

165. Professor Erich Israel Lehmann, Sterberegister der Berliner Standesämter 1874–1955 (Death Register of the Berlin Registry Offices 1874–1955), Landesarchiv, Berlin, Deutschland.

Chapter 3: Physics and Chemistry Survivors

166. R. E. Wolman, *Crossing over. An Oral History of Refugees from Hitler's Reich* (Twayne Publishers, New York, 1996).

167. Dreyfuss, Paul (1933–1934, 1938–1939), Box 52.56, Emergency Committee in Aid of Displaced Foreign Scholars Records, New York Public Library.

168. Paul Dreyfuss, Foreigners' File No. A271.228, National Archives of Belgium.

169. N. Daugaard, Avant-gardist colors in a political tug-of-war. Gasparcolor between art and fascism, in B. Flückiger, E. Hielscher and N. Wietlisbach (Eds.) *Color Mania — the Material of Color in Photography and Film*, pp. 187–195 (Lars Müller, Zurich, 2020).

170. W. Moritz, *Gaspar Color: Perfect Hues for Animation*, www.oskarfischinger.org/GasparColor.htm.

171. Private communication Michael Dreyfuss and Judy Navon-Dreyfuss.

172. J. M. Deem, *The Prisoners of Breendonk. Personal Histories from a World War II Concentration Camp* (Houghton Mifflin Harcourt, New York, 2015).

173. W. Michaelis, *Process for Producing Photographic Multicolor Pictures*, 15 November 1940 (USA Patent 2,347,119).

174. R. C. Miller, *Frozen Moments*, M. Andrews and R. Vogel (Eds.) (Bombshelter Press, Los Angeles, 2009).

175. B. Gaspar and P. D. Dreyfuss, *Acid Azo Dyes*, 2 April 1948 (USA Patent 2,612,496).

176. Cibachrome Prints: Elusive and Beautiful, www.lumieregallery.net/13483/cibachrome-prints-elusive-and-beautiful/.

177. V. Henri, Absorption spectra of polyatomic molecules. Predissociation and dissociation of these molecules. *Trans. Faraday Soc.*, **25**, 765 (1929).

178. Paul Goldfinger, Foreigners' File No. A87.144, National Archives of Belgium.

179. Einstein to Ettlinger, 4 May 1933, www.livinghistoryofillinois.com/pdf_files/Abraham-Lincoln-Forgeries-Joseph-Cosey.pdf.

180. Lasareff, Dr. Vladimir (1904-), File 1935–47, MS. S.P.S.L. 333/9.

181. P. Goldfinger, V. Lasareff and B. Rosen, L'Energie de Dissociation de l'Oxyde de Carbone (The dissociation energy of carbon monoxide). *Compt. Rend. Hebd. Séances Acad. Sci.*, **201**, 958 (1935).

182. V. Lasareff and J. Roskam, Quelques Techniques Permettant d'Observer, au Niveau d'une Surface Solide, la Présence de Protéines Absorbées (Some techniques permitting the observation, at the level of a solid surface, the presence of absorbed proteins). *Compt. Rend. Hebd. Séances Soc. Biol. Fil.*, **132**, 479 (1939).

183. V. Lasareff, *La Vie Remporta la Victoire* (*Life Won Victory*) (Seine et Meuse, Liège/Paris/Genève, 1945).

184. P. Nefors, *Breendonk 1940–1945* (Standaard, Antwerp, 2005).

185. Einstein to Ettlinger, 16 December 1944, ilab.org/assets/catalogues/catalogs_files_3119_catalogue_0_galerie_thomas_vincent.pdf.

186. Einstein to Ettlinger, 15 February 1945, christies.com/en/lot/lot-6296879.

187. V. Lasareff, Mesure du pH au Moyen d'un Amplificateur Balistique, (pH measurement by means of a Ballistic Amplifier). *Bull. Soc. Chim. Belg.*, **56**, 36 (1947).

188. Utilization of Atomic Energy Scientific and Technical Information: Minutes of the International Conference Held in Geneva, Switzerland,

Sponsored by the United States Atomic Energy Commission, May 26–29, 1958.

189. World Nuclear Directory (Harrap, London, 1960).

190. C. P. Enz, *No Time to Be Brief: A Scientific Biography of Wolfgang Pauli* (Oxford University Press, 2002).

191. P. Goldfinger and V. Lasareff, La Réaction des Amines avec L'Eau Lourde (The reactions of amines with heavy water). *Compt. Rend. Hebd. Séances Acad. Sci.*, **200**, 1671 (1935).

192. Goldfinger, Dr. Paul (1905–70), File 1935–47, MS. S.P.S.L. 213/6.

193. J. B. Tucker, *War of Nerves. Chemical Warfare from World War I to Al-Qaeda* (Pantheon Books, New York, 2006).

194. Marianne Goldfinger, interview by Cathy Courtney, 22 December 2002, NLSC: Architects' Lives, British Library Sound Archive, C467/74/01-04.

195. J. Duchesne, P. Goldfinger and B. Rosen, Heat of atomisation of carbon. *Nature*, **159**, 130 (1947).

196. P. Goldfinger, *Color Photographic Material*, 23 February 1939 (USA Patent 2,283,361).

197. Peter Goldfinger, interview by Cathy Courtney, 1 September 1999, NLSC: Architects' Lives, British Library Sound Archive, C467/44/01-14.

198. P. Goldfinger, R. M. Noyes and W. Y Wen, Gas phase chlorine plus hydrogen bromide reaction. A bimolecular reaction of diatomic molecules. *J. Am. Chem. Soc.*, **91**, 4003 (1969).

199. P. Goldfinger and G. Verhaegen, Stability of the gaseous ammonium chloride molecule. *J. Chem. Phys.*, **50**, 1467 (1969).

200. P. Goldfinger, Mass spectroscopy of inorganic system at high temperatures. *Angew. Chemie*, **3**, 153 (1964).

201. A. D. Vergallo and E. L. Wetzlar, Yves Cape, AFC, SBC, Wiping the Slate Clean, *French Association of Directors of Cinematographic Photography*, 17 April 2018.

202. I. Fleming, *Goldfinger* (The Book Club, London, 1959).

203. Rosen, Dr. Boris (1900-), File 1935-46, MS. S.P.S.L. 338/2.

204. Boris Rosen, Foreigners' File No. A84.104, National Archives of Belgium.

205. B. Rosen, *Données Spectroscopiques Concernant les Molécules Diatomiques* (*Spectroscopic Data Relative to Diatomic Molecules*) (Hermann, Paris, 1951).

206. Half a Century of Space Research at the University of Liège, www.academieairespace.com/wp-content/uploads/2018/05/rocchus_bxl.pdf.

207. B. Rosen and J. Depireux (Eds.) *Optical Spectroscopy of Solids. Proceedings of the Xth European Congress on Molecular Spectroscopy,* University of Liège, September 29-October 4, 1969.

208. A. Lustig and M. Reiss, Kritische Bemerkungen zur Auswertung der Ladungsmessungen an Kleinen Probekörpern (Critical comments on the evaluation of the charge measurements on small specimens). *Z. Phys.*, **84**, 131 (1933).

209. Lustig Dr. Alfred (1908-), File 1938–48, MS. S.P.S.L. 334/6.

210. Alfred Lustig, Austria, Vienna, Jewish Emigrant Applications, 1938–1939, www.myheritage.com.

211. G. Schneider, *Exile and Destruction. The Fate of Austrian Jews 1938–1945* (Praeger, Westport CT, 1995).

212. Alfred Lustig, Private Biography, sent to the author by Lustig's colleague Professor M. F. Gilmullin (Elabuga Institute of Kazan Federal University, Russia).

213. P. Zusmanovich, Mathematicians Going East, arXiv: 1805.00242 (math), 2018.

214. M. F. Gilmullin, *Alfred Lustig Austrian Scientist in Yelabuga*, www.nwm.at/vse-ob-avstrii/znamenitye-avstrijtsy/alfred-lyustig-avstrijskij-uchenyj-v-elabuge.

215. Alfred Lustig in Memorial Book for the Victims of National Socialism at the University of Vienna, gedenkbuch.univie.ac.at.

216. L. Schiavone, Oltre l'Astronomia, la Vita: Giulio Bemporad e L'Assistenza ai Profughi Ebrei (Beyond astronomy, life: Giulio Bemporad and assistance to Jewish refugees). *G. Astronom.*, **2**, 15 (2015).

217. M. A. Livingston, *The Fascists and the Jews of Italy. Mussolini's Race Laws, 1938–1943* (Cambridge University Press, 2014).

218. D. N. Schwartz, *The Last Man Who Knew Everything, The Life and Times of Enrico Fermi, Father of the Nuclear Age* (Basic Books, New York, 2017).

219. A. Capristo, The Persecution and Emigration of Jewish Mathematicians, Astronomers and Physicists: The Case of Fascist Italy, in L. Saraiva (Ed.) *Proceedings of the International Conference Mathematical Sciences under Dictatorships: Western Europe, Portugal, and Its Atlantic Connections,* Faculty of Sciences of Lisbon University, December 10–12, 2015, pp. 83–122 (Sociedade Portuguesa de Matemática, Lisbon, 2018).

220. R. Nossum, Emigration of mathematicians from outside German-speaking academia 1933–1963, supported by the Society for the Protection of Science and Learning. *Hist. Math.*, **39**, 84 (2012).

221. Fano, Professor Gino (1871-), File 1939–47, MS. S.P.S.L. 278/6.

222. Fano, Dr. Ugo (1912-), File 1938–48, MS. S.P.S.L. 327/3.

223. Segre, Professor Beniamino (1903–1977), File 1938–60, MS. S.P.S.L. 285/1.

224. Bemporad, Professor Giulio (1888-), File 1938–39, MS. S.P.S.L. 323/6.

225. L. Volta, Giulio Bemporad, *Memorie Della Società Astronomica Italiana*, **19**, 227 (1948).

226. M. von Laue and K. H. Riewe, Der Kristallformfaktor für das Oktaeder (The crystal formation factor for the octahedron). *Z. Kristall.*, **95**, 408 (1936).

227. Riewe, Dr. Karl-Heinrich (1901-), File 1935–43, MS. S.P.S.L. 337/8.

228. F. G. Houtermans and K. H. Riewe, Über die Raumladungswirkung an Einem Strahl Geladener Teilchen von Rechteckigem Querschnitt der Blende (About the space charge effect on a beam of charged particles with a rectangular cross-section of the diaphragm). *Archiv Elektrotechnik*, **35**, 686 (1941).

229. Houtermans, Professor Fritz Georg (1903–1966), File to 1950, MS. S.P.S.L. 330/8.

230. M. Shifman, *Physics in a Mad World* (Based on the writings of V. J. Frenkel) (World Scientific Publishing, Singapore, 2015).

231. E. Amaldi, *The Adventurous Life of Friedrich Georg Houtermans, Physicist (1903–1966)*, S. Braccini, A. Eredidato and P. Scampoli (Eds.) (Springer, Heidelberg, 2012).

232. G. Gamow and F. G. Houtermans, Zur Quantummechanik des Radioaktiven Kerns (On the quantum mechanics of the radioactive nucleus). *Z. Phys.*, **52**, 496 (1928).

233. R. d'E. Atkinson and F. G. Houtermans, Transmutation of the lighter elements in stars. *Nature*, **123**, 567 (1929).

234. Bethe, Dr. Hans (1906-), File 1934-45, MS. S.P.S.L. 324/4.

235. O. R. Frisch, *What Little I Remember* (Cambridge University Press, 1979).

236. I. B. Khriplovich, The eventful life of Fritz Houtermans. *Physics Today*, **45**, 29 (1992).

237. F. G. Houtermans, A. I. Leipunsky and L. Rusinow, The absorption of group C-neutrons in silver, cadmium and boron at different temperatures. *Phys. Zeit. Sowjetunion*, **12**, 491 (1937).

238. H. Hellmann, *Einführung in die Quantenchemie* (*Introduction to Quantum Chemistry*) (Franz Deuticke, Leipzig, 1937).

239. R. P. Feynman, Forces in molecules. *Phys. Rev.*, **56**, 340 (1939).

240. W. H. E. Schwarz, D. Andrae, S. R. Arnold, J. Heidberg, H. Hellmann Jr., J. Hinze, A. Karachalios, M. A. Kovner, P. C. Schmidt and L. Zülicke, Hans G. A. Hellmann (1903–1938) a pioneer of quantum chemistry. *Bunsen-Magazin* (1 and 2) 10–21 and 60–70 (1999).

241. M. Healea and C. Houtermans, The relative secondary electron emission due to He, Ne, and Ar ions bombarding a hot nickel target. *Phys. Rev.*, **58**, 608 (1940).

242. M. Shifman (Ed.) *Standing Together in Troubled Times. Unpublished Letters by Pauli, Einstein, Franck and Others* (World Scientific Publishing, Singapore, 2017).

243. A. Weissberg, *Conspiracy of Silence* (Hamish Hamilton, London, 1952).

244. Manfred Von Ardenne, (1907–1997), www.vonardenne.biz/en/company/mva/.

245. Thirring to Von Ardenne, 15 July 1942, Phaidra Archiv, Österreichische Zentralbibliothek für Physik, Wien.

246. F. G. Houtermans, Zur Frage der Auslösung von Kern-Kettenreaktionen (On the question of triggering nuclear chain reactions), unpublished report from laboratory of M. von Ardenne, Berlin-Lichterfelde-Ost, Germany, August 1941, G 94 in microfilm archive of captured German documents, Natl. Tech. Inf. Service, Oak Ridge, Tenn., USA.

247. D. C. Cassidy, *Beyond Uncertainty: Heisenberg, Quantum Physics, and the Bomb* (Bellevue Literary Press, New York, 2010).

248. A. Wolff, A conversation with Nobel prize winner Eugene P. Wigner. *Look*, **12**, 56 (1967).

249. Contributions of German Scientists to the Soviet Atomic Energy Program Elektrostal, CIA Scientific Intelligence Report 2-RS IV-57, 15 July 1957 (Approved for release 12/03/2018), The National Security Archive, Gelman Library, The George Washington University.

250. N. Riehl, *Zehn Jahre im Goldenen Käfig. Erlebnisse beim Aufbau der Sowjetischen Uran-Industrie* (*Ten Years in a Golden Cage. Experiences in Building the Soviet Uranium Industry*) (Riederer-Verlag, Stuttgart, 1988).

251. P. V. Oleynikov, German scientists in the soviet atomic project. *Non-Proliferation Rev.*, **7**, 1 (2000).

252. Interview with K.-H. Riewe, *Reports from Prisoners Repatriated from Prison Camps in Soviet Union: Appeals on Behalf of Some Internees*, FO 371/122936 (1956), UK National Archives Kew.

253. S. A. Barnes, 'In a manner befitting Soviet citizens': An uprising in the post-Stalin Gulag. *Slavic Rev.*, **64**, 823 (2005).

254. A. Soltzenhitsyn, *The Gulag Archipelago* (Harvill Press, London, 1995).

255. J. Treusch and F. Dreisigacker, Dank an Karl-Heinrich Riewe (Thanks to Karl-Heinrich Riewe). *Physikalische Blätter*, **40**, 386 (1984).

256. E. Brüche, K.-H. Riewe 70 Jahre. *Physikalische Blätter*, **33**, 363 (1977).

Chapter 4: Top-Secret Refugees

257. Frisch, Professor Otto Robert (1904–1979), File 1937-48, MS. S.P.S.L. 327/10.

258. L. Meitner and O. R. Frisch, Products of the fission of the uranium nucleus. *Nature*, **143**, 471 (1939).

259. R. W. Clark, *Tizard* (MIT Press, Massachusetts, 1965).

260. P. Gillman and L. Gillman, *'Collar the Lot!' How Britain Interned and Expelled Its Wartime Refugees* (Quartet Books, London, 1980).

261. R. Peierls, Otto Robert Frisch. 1 October 1904–22 September 1979. *Biogr. Mems. Fell. R. Soc.*, **27**, 283 (1981).

262. Fuchs, Dr. Emil Klaus Julius (1911–1988), File 1937-51, MS. S.P.S.L. 328/2.

263. N. T. Greenspan, *Atomic Spy: The Dark Lives of Klaus Fuchs* (Viking, New York City, 2020).

264. F. Close, *Trinity: The Treachery and Pursuit of the Most Dangerous Spy in History* (Penguin, London, 2019).

Chapter 5: Refugees in Mathematics

265. Heilbronn, Professor Hans Arnold (1908–1975), File 1933-76, MS. S.P.S.L. 279/7.

266. J. W. S. Cassels and A. Fröhlich, Hans Arnold Heilbronn. 8 October 1908–28 April 1975. *Biogr. Mems Fell. R. Soc.*, **22**, 119 (1976).

267. R. T. Nossum and J. Kotůlek, The Society for the Protection of Science and Learning as a patron of refugee mathematicians. *BSHM Bull. J. Brit. Soc. Hist. Math.*, **30**, 153 (2015).

268. C. R. Fletcher, Refugee mathematicians: A German crisis and a British response, 1933–1936. *Hist. Math.*, **13**, 13 (1986).

269. M. Pinl, In Memory of Ludwig Berwald. *Scripta Math.*, **XXVII**, 193 (1964).

270. Berwald, Professor Ludwig (1883–1939), File 1939, MS. S.P.S.L. 470/1.

271. L. Berwald, Uber Finslersche und Cartansche Geometrie. IV. Projektivkrummung Allgemeiner Affiner Räume und Finslersche Räume Skalarer Krümmung (On Finsler's and Cartan's Geometry. IV. Projective Curvature of General Affine Spaces and Finslerian Spaces of Scalar Curvature). *Ann. Math.*, **48**, 755 (1947).

272. J. Kotůlek and R. T. Nossum, Jewish Mathematicians Facing the Nazi-Threat: The Case of Walter Fröhlich. *Jud. Bohemiae*, **48**, 69 (2013).

273. Fröhlich, Dr. Walter (1902), File 1939-48, MS. S.P.S.L. 489/1.

274. J. Buresova, The Czech Refugee Trust Fund in Britain 1939–1950 in *Exile in and from Czechoslovakia During the 1930s and 1940s*, pp. 133–145, C. Brinson and M. Malet (Eds.) (Rodopi, Amsterdam, 2009).

275. Löwner, Professor Karl (1893–1968), File 1939-47, MS. S.P.S.L. 282/2.

276. J. J. O'Connor and E. F. Robertson, Charles Loewner, mathshistory.st-andrews.ac.uk/Biographies/Loewner/.

277. Erdelyi, Dr. Arthur (1908–1977), File 1938-78, MS. S.P.S.L. 278/4.

278. W. Parys, Why didn't Charasoff and Remak use Perron-Frobenius mathematics? *Eur. J. Hist. Econ. Thought*, **21**, 991 (2014).

279. Remak, Dr. Robert Erich (1888–c.1945), File 1934-46, MS. S.P.S.L. 540/3.

280. R. Siegmund-Schultze, *Mathematicians Fleeing from Nazi Germany, Individual Fates and Global Impact* (Princeton University Press, 2009).

281. L. Brockliss, Welcoming and Supporting Refugee Scholars: The Role of Oxford's Colleges, in *Ark of Civilization: Refugee Scholars and Oxford University, 1930–1945*, S. Crawford, K. Ulmschneider and J. Elsner (Eds.), pp. 62–76 (Oxford University Press, 2017).

282. J. J. O'Connor and E. F. Robertson, Robert Erich Remak, mathshistory. st-andrews.ac.uk/Biographies/Remak/.

283. Amsterdam Police Reports 1940–1945, Part: 6303, Period: 1803–1956, No. 5225, Politirapporten '40-'45, Amsterdam City Archives.

284. Dutch Jewish Genealogical Database, www.dutchjewry.org/inmemoriam/r.shtml.

285. P. Butzer and L. Volkmann, Otto Blumenthal (1876–1944) in Retrospect. *J. Approx. Th.*, **138**, 1 (2006).

286. H. T. Jongen and A. Krieg, Otto Blumenthal. *DMV-Mitteilungen*, **4**, 49 (2000).

287. Blumenthal, Otto (1933–1935), Box 44.55, Emergency Committee in Aid of Displaced Foreign Scholars Records, New York Public Library.

288. Blumenthal, Professor Otto (1876–1944), File 1934-47, MS. S.P.S.L. 471/1.

289. K. P. E. Gravemeijer and J. Terwel, Hans Freudenthal: A mathematician on didactics and curriculum theory. *J. Curriculum. Stud.*, **32**, 777 (2000).

290. Gedenktafel für Otto Blumenthal in Aachen (Deutschland) (Commemorative Plaque for Otto Blumenthal in Aachen (Germany)), www.w-volk.de/museum/plaqu100.htm.

291. Funk, Dr. Paul (1886–1969), File 1939, MS. S.P.S.L. 279/2.

292. N. Schappacher, The Nazi Era: The Berlin Way of Politicizing Mathematics, in *Mathematics in Berlin*, pp. 127–136, H. Begehr, H. Koch, J. Kramer, N. Schappacher and E.-J. Thiele, (Eds.), (Birkhäuser Verlag GmbH, Berlin, 1998).

293. L. Vastenhout, *Between Community and Collaboration: 'Jewish Councils' in Western Europe under Nazi Occupation* (Cambridge University Press, 2022).

294. M. R. Marrus and R.O. Paxton, The Nazis and the Jews in Occupied Western Europe, 1940–1944. *J. Mod. Hist.*, **54**, 687(1982).

295. L. de Jong, *The Netherlands and Nazi Germany* (Harvard University Press, 1990).

Chapter 6: Refugees in Medicine

296. P. Weindling, Medical refugees and the modernisation of British medicine, 1930–1960. *Soc. Hist. Med.*, **22**, 489 (2009).

297. D. Pyke, Contributions by German émigrés to British medical science. *Q. J. Med.*, **93**, 487 (2000)

298. Medical Sub-Series, MS. S.P.S.L., 358–425.

299. L. A. Zeidman and D. Kondziella, Neuroscience in Nazi Europe Part III: Victims of the Third Reich. *Can. J. Neurol. Sci.*, **39**, 729 (2012).

300. A. Simons, Kopfhaltung und Muskeltonus Klinische Beobachtungen (Head posture and muscle tone clinical observations). *Z. Neur. Psych.*, **80**, 499 (1923).

301. B. Holdorff, Arthur Simons (1877–1942) and tonic neck reflexes with hemiplegic 'Mitbewegungen' (associated reactions): Cinematography from 1916–1919. *J. Hist. Neurosci.*, **25**, 63 (2016).

302. Simons, Professor Arthur (1877–c.1945), File 1933-46, MS. S.P.S.L. 551/3.

303. G. J. Fraenkel, *Hugh Cairns: First Nuffield Professor of Surgery, University of Oxford* (Oxford University Press, 1991).

304. M. Kingreen and W. Scheffler, The deportations to Raasiku near Reval, in *Buch der Erinnerung, Die ins Baltikum Deportierten Deutschen, Österreichischen und Tschechoslowakischen Juden* (*The German, Austrian and Czechoslovakian Jews Deported to the Baltic States*), Volksbund Deutsche Kriegsgräberfürsorge e.V. and Riga-Komitee der Deutschen Städte (Eds.), pp. 869–914 (K. G. Saur, Berlin, 2003).

305. Commemoration at Kalevi-Liiva, holocaustremembrance.com/news-archive/commemoration-kalevi-liiva.

306. Sittig, Professor Otto (1886–1945), File 1939-43, MS. S.P.S.L. 552/2.

307. Guttmann, Sir Ludwig (1899–1980), File 1938-80, MS. S.P.S.L. 394/8.

308. D. Whitteridge, Ludwig Guttmann, 3 July 1899–18 March 1980. *Biogr. Mems Fell. R. Soc.*, **29**, 226 (1983).

309. A. Compston, Sir Henry Head FRS (1861–1940): A life in science and society. *J. Neurol. Neurosurg. Psychiatry*, **88**, 716 (2017).

310. O. Sittig and J. Urban, Case of poliomyelitis with bilateral paralysis of masticatory muscles. *Lancet*, **1**, 865 (1939).

311. O. Sittig, *Aryanization and Confiscation of Property in the Protectorate*, RG-48.008M, Selected Record from Central State Archives in Prague (Fond JAF 1005).

312. A. Autenrieth, *Ärztinnen und Ärzte am Dr. von Haunerschen Kinderspital, die Opfer Nationalsozialistischer Verfolgung Wurden* (*Doctors at Dr. von Hauner Children's Hospital Who Became Victims of National Socialist Persecution*), Dissertation, Fakultät der Ludwig-Maximilians-Universität zu München (2012).

313. L. P. Johannsen, *Erich Aschenheim, Albert Eckstein and Julius Weyl, Jüdische Pädiater im Vorstand der Vereinigung Rheinisch-Westfälischer Kinderärzte* (*Jewish Pediatricians on the Board of the Association of Pediatricians in Rhineland-Westphalia*) (Hentrich & Hentrich, Berlin, 2010).

314. Aschenheim, Dr. Erich (1882-), File 1936-46, MS. S.P.S.L. 466/2.

315. Tracing Request for Aschenheim, Erich (1882-11-04), Doc. ID. 85985465, ITS Digital Archive, Arolsen Archives.

316. A. Grau, Geschichte von Planegg im 19. und 20. Jahrhundert, in *Planegg, Geschichte und Geschichten (*History of Planegg in the 19th and 20th centuries, in *Planegg, History and Stories*) Community Planegg, Community Archive Krailling, **II** (2009).

317. F. Kraus, Ferdinand Blumenthal, *Z. Krebsforschung,* **32**, 2 (1930).

318. H. Jenss and P. Reinicke, *Ferdinand Blumenthal: Kämpfer für eine Fortschrittliche Krebsmedizin und Krebsfürsorge (Ferdinand Blumenthal: Fighter for Progressive Cancer Medicine and Cancer Care)* (Hentrich & Hentrich, Berlin, 2012).

319. More Wholesale Dismissals of World Famous Scholars Ordered by Nazi Minister, *Jewish Daily Bulletin*, New York, May 3, 1933.

320. M. Samardžić and M. Bešlin, The Work of German oncologist Ferdinand Blumenthal in the Kingdom of Yugoslavia, 1933–1937. *Vojnosanit. Pregl.*, **76**, 847 (2019).

321. Blumenthal, Professor Ferdinand (1868–1941), File 1938-44, MS. S.P.S.L. 378/9.

322. Ferdinand Blumenthal, Charité Memorial Site, Universitätsmedizin Berlin, gedenkort.charite.de/en/people/ferdinand_blumenthal/.

323. P. Voswinckel, In Memoriam Hans Hirschfeld (1873–1944). *Folia Haematol.*, *Leipzig*, **114**, 707 (1987).

324. H. Hirschfeld and A. Hittmair, *Handbuch der allgemeinen Hämatologie (Handbook of General Haematology)* (Urban and Scharwzenberg, Berlin, 1932).

325. Hirschfeld, Professor Hans (1873-[UNK]), File 1938-43, MS. S.P.S.L. 501/2.

326. P. Voswinckel, 'Verweigerte Ehre' Dokumentation zu Hans Hirschfeld, 1937–2012, Die Geschichte der Deutschen Gesellschaft für Hämatologie und Onkologie im Spiegel ihrer Ehrenmitglieder ('Denied Honour' Documentation on Hans Hirschfeld, 1937–2012, The History of the German Society for Haematology and Oncology as Reflected by its Honorary Members) (Deutsche Gesellschaft für Hämatologie und Onkologie, Berlin 2012).

327. L. Heilmeyer and A. Hittmair, *Handbuch der gesamten Hämatologie (Handbook of the Whole of Haematology)* (Urban and Scharwzenberg, Berlin, 1957).

328. S. Laudien, Abhängung der Gedenktafel für Ludwig Heilmeyer (Removal of the Memorial Plaque for Ludwig Heilmeyer), University of Jena, 18 May 2021, uni-jena.de/210518-heilmeyer.

329. F. Steger and J. Jeskow, *Ludwig Heilmeyer. A Political Biography* (Franz Steiner Verlag, Stuttgart, 2021).

330. Südwest Presse, Strassenname: Die Causa Heilmeyer, Ausgabe Ulm/Neu-Ulm vom 23 November 2017.

331. Belated tribute to Jewish-born haematologist. City of Ulm and Ulm University dedicate Hans Hirschfeld Square, uni-ulm.de/en/university/ alumni/downloads/newsletter-archive/alumni-news-03-2021/.

332. Kral, Adalbert (1903-), File 1939-43, MS. S.P.S.L. 396/2.

333. Vojtech Adalbert Kral, Professor Emeritus of Clinical Psychiatry, University of Western Ontario, Canada. *Bull. R. Coll. Psychiatrists*, **12**, 395 (1988).

Chapter 7: Refugees in Biology

334. Biology, MSS. S.P.S.L. 195–207.

335. P. Weindling, Britain's role supporting refugee biologists escaping war and persecution. *The Biologist*, **61**, 24 (2023).

336. Simons, Dr. Hellmuth (1893-), File 1933-47, MS. S.P.S.L. 205/2.

337. H. Simons, Saprophytische Oscillarien des Menschen und der Tiere (Saprophytic oscillaria of humans and animals). *Zbl. Bakt. (1 Abt. Orig.)*, **88**, 501 (1922).

338. T. R. R. Mann, David Keilin 1887–1963. *Biogr. Mems Fell. R. Soc.*, **10**, 182 (1964).

339. Simons, Dr. Helmuth Carl Rudolf, Lehmann-Russbueldt, Dr. Otto, *Security Services: Personal (PF Series) Files, Communist and Suspected Communists, Including Russian and Communist Sympathisers*, KV 2/2001-6, UK National Archives Kew.

340. H. W. Steed, Aerial warfare: Secret German plans. *The Nineteenth Century and After*, **116**, 1 (1934).

341. O. Lehmann-Russbüldt, *Hitler's Wings of Death* (Telegraph Press, New York, 1936).

342. H. Liepmann, *Death From The Skies — A Study of Gas and Microbial Warfare* (Martin Secker & Warburg, London, 1937).

343. M. J. Trow, *The Black Book* (John Blake, London, 2017).

344. Simons, Hellmuth Karl Rudolf, Police File N09007, Swiss Federal Archives, Federal Department of Home Affairs, Berne.

345. H. C. R. Simons, *Escape from Death in France*, Wiener Library 1656/3/9/60.

346. Simons, Hellmuth, German and Jewish Intellectual Émigré Collections, American Council for Émigrés in the Professions Records, 1930–1974, Files from Else Staudinger, Folder 126, Special Collections and Archives, University at Albany.

347. Allen Dulles, US Office of Strategic Services to Washington, Telegram 1244–45, 8 December 1943, in N. H. Petersen, *From Hitler's Doorstep:*

The Wartime Intelligence Reports of Allen Dulles, 1942–1945 (Pennsylvania State University Press, 1996).

348. E. Croddy, C. Perez-Armendariz and J. Hart, *Chemical and Biological Warfare. A Comprehensive Survey for the Concerned Citizen* (Copernicus Books, New York, 2002).

349. L. Tatu and J.-P. Feugeas, Botulinum toxin in WW2 German and allied armies: Failures and myths of weaponization. *Eur. Neurol.*, **84**, 53 (2021).

350. Hellmuth Simons, Correspondence, in Guggenheim, Alis (1896–1958) Archive, Ref. Ar 126.5, Schweizerisches Sozialarchiv. See also: H. Holz, S. Gisel-Pfankuch, U. Hobi and B. Wismer, *Alis Guggenheim (1896–1958)* (Verlag Lars Müller, Baden, 1992).

351. Rapid Detection of Blood Infections. *Swiss Society for Research Expeditions*, 31 October 1946.

352. H. C. R. Simons, *Science*, **104**, 465 (1946).

353. Refugee Scientist is Studying Polio. *Pennsylvania Daily News*, 19 August 1947.

354. H. C. R. Simons, Ist die Multiple Sklerose eine Spirochätose? (Is disseminated sclerosis caused by spirochaetes). *Deutsche Medizin Wochenschrift*, **83**, 1196 (1958).

355. Werner Simons (1920–1994), Ancestry, ancestry.co.uk.

356. Tchernavin, Dr. Vladimir (1887–1949), File 1936-47, MS. S.P.S.L. 206/1.

357. T. Tchernavin, *Escape from the Soviets* (Tr. N. Alexander) (E.P. Dutton, New York, 1934).

358. V. V. Tchernavin, *I Speak for the Silent Prisoners of the Soviets* (Tr. N. M. Oushakoff) (Hale, Cushman and Flint, New York, 1935).

359. V. B. Wigglesworth, Boris Petrovitch Uvarov (1889–1970). *Biog. Mems Fell. R. Soc.*, **17**, 713 (1971).

360. Rutherford of Nelson, The Society for the Protection of Science and Learning. *Science*, **83**, 372 (1936).

361. A. Einstein, E. Schrödinger and V. Tchernavin, The freedom of learning. *Science*, **83**, 372 (1936).

362. Portrait of the artist as a director. *Harvard Magazine*, September-October 2002.

363. V. Tchernavin, The breeding characters of salmon in relation to their size. *Proc. Zoo. Soc.*, **B113**, 206 (1944).

364. N. Coward and D. Lean, *In Which We Serve*, www.imdb.com/title/tt0034891/.

365. V. Tchernavin, A living bony fish which differs substantially from all living and fossil Osteichthyes. *Nature*, **158**, 667 (1946).

366. Professor's Death. *The Times*, 5 April 1949.

367. E. Trewavas, Dr. Vladimir Tchernavin. *Nature*, **163**, 755 (1949).

368. M. Y. Sorokina, Within two tyrannies: The Soviet academic refugees of the Second World War, Chapter 14, in S. Marks, P. Weindling and L. Wintour (Eds.) *In Defence of Learning: The Plight, Persecution and Placement of Academic Refugees, 1933–1980s* (Oxford University Press for the British Academy, 2011).

369. Engineer's Heroic Escape is Subject of a TV Documentary. *New Civil Engineer*, 15 July 1999.

Chapter 8: Refugees in Engineering

370. B. Pistole and B. Meyer, Alfred Rheinheimer, February 2022, www. stolpersteine-hamburg.de.

371. Rheinheimer, Dr. Alfred (1884-), File 1933-46, MS. S.P.S.L. 245/8.

372. Alfred Rheinheimer, First World War Internee, Knockaloe Camp, Isle of Man, Prisoner of War Information Bureau Serial No. 15321, www. imuseum.im/search/collections/people/mnh-agent-1155933.html.

373. 75.000 'Stolpersteine' Erinnern an Nazi-Opfer (75,000 'Stumbling Blocks' Commemorate Nazi Victims), *Deutsche Welle*, 29 December 2019.

Chapter 9: Refugees in Social Sciences

374. C. Fleck, Austrian refugee social scientists, Chapter 12, in S. Marks, P. Weindling and L. Wintour (Eds.) *In Defence of Learning: The Plight, Persecution and Placement of Academic Refugees, 1933–1980s* (Oxford University Press for the British Academy, 2011).

375. M. Lippman, They shoot lawyers don't they? Law in the Third Reich and the global threat to the independence of the judiciary. *Cal. West. Int. Law J.*, **23**, 257 (1993).

376. J. Friedlander, *A Light in Dark Times: The New School for Social Research and Its University in Exile* (Columbia University Press, 2019).

377. Waldecker, Dr. Ludwig (1881-), MS. S.P.S.L. No file number.

378. Mendelssohn-Bartholdy, Professor Albrecht (1874–1936), File 1933-37, MS. S.P.S.L. 525/3.

379. Cohn, Professor Ernst Josef (1904–1976), File 1933-75, MS. S.P.S.L. 262/9.
380. B. Collins, By post or by ghost: Ruminations on visions and epistolary archives. *J. Quart. Rev.*, **107**, 397 (2017).
381. B. Collins, *Robert Eisler and the Magic of the Combinatory Mind: The Forgotten Life of a 20th Century Austrian Polymath* (Palgrave Macmillan, New York, 2021).
382. R. Eisler, *Studien zur Werttheorie* (*Studies in the Theory of Value*) (Duncker and Humblot, Leipzig, 1902).
383. R. Eisler, *Das Geld, Seine Geschichtliche Entstehung und Gesellschaftliche Bedeutung: ein Wirtschaftswissenschaftlicher Lichtbild-Lehrgang* (*Money, Its Historical Origin and Social Significance, an Economics Course*) (Diatypie, Munich, 1924).
384. R. Eisler, *The Enigma of the Fourth Gospel* (Methuen, London, 1938).
385. R. Eisler, The empiric basis of moral obligation. *Ethics*, **LIX**, 77 (1949).
386. Einstein to Eisler, 31 January 1925, Vol. 14: The Berlin Years, Writings and Correspondence, April 1923-May 1925, No. 428, *The Collected Papers of Albert Einstein* (Princeton Press, 1987).
387. Einstein to Pribram, 1 May 1926, Lot 515 — A178 Manuscripts & Autographs, www.kollerauktionen.ch/en/home.htm.
388. R. Eisler, *The Money Maze. The Way Out of the Economic Crisis* (Search Publishing, London, 1931).
389. R. Eisler, *Stable Money: The Remedy for the Economic World Crisis* (Search Publishing, London, 1932).
390. Eisler, Dr. Robert (1892–1949), File 1939, MS. S.P.S.L. 230/3.
391. B. Collins, *A Very Square Peg. The Strange and Remarkable Life of the Polymath Robert Eisler*, Podcast, Episode 9: Vanity of Vanities (New Books Network, 2020).
392. Otto Eisler, *The Orpheus Trust*, klangwege.orpheustrust.at/musikschaffende_e.php?detail=9.
393. R. Eisler, *The Royal Art of Astrology* (Herbert Joseph, London, 1946).
394. L. Anderson, *The Diaries* (Methuen, London, 2004).
395. Robert Eisler, Dr., Letter to *The Times* on Goethe's Birthplace, 7 April 1945.
396. Robert Eisler, Dr., Letter to *The Times* on Bolivar in London, 8 January 1948.
397. Robert Eisler, Dr., Letter to *The Times* on the Obituary of the Führer, 7 May 1945.
398. R. Eisler, *Man into Wolf* (Spring Books, London, 1948).

399. P. Vronsky, *American Serial Killers. The Epidemic Years 1950–2000* (Berkley, New York, 2021).

400. An Encyclopaedic Author, Dr. Robert Eisler, Obituary, *The Times*, 20 December 1949.

401. C.J. Goldsmid, Capt., Letter to *The Times* on Dr. Robert Eisler, 18 January 1950.

402. Lili (Rosalia) von Pausinger, www.myheritage.com.

403. O. Fröbe-Kapteyn, *Eranos-Jahrbuch, 1946, Band XIV: 'Geist und Natur' (Spirit and Nature)* (Rhein-Verlag, Zurich, 1946).

404. B. Collins, *A Very Square Peg. The Strange and Remarkable Life of the Polymath Robert Eisler*, Podcast, Episodes 1–9 (New Books Network, 2020).

405. H. Hagemann, Wissenschaftliche Würdigung von Paul Maximilian Eppstein (1902–1944) (Academic appraisal of Paul Eppstein), *Mannheimer Geschichtsblätter*, **39**, 3 (2020).

406. Eppstein, Dr. Paul Maximilian (1902-), File 1933-48, MS. S.P.S.L. 230/6.

407. Testimony of Benno Cohn about His Meeting With Eichmann, March 1939, *The Trial of Adolf Eichmann*, Vols. I–V, State of Israel, Ministry of Justice, Jerusalem, 1994.

408. B. Prager, Interpreting the visible traces of Theresienstadt, *J. Mod. Jewish Stud.*, **7**, 175 (2008).

409. L. Rothkirchen, *The Jews of Bohemia and Moravia: Facing the Holocaust* (University of Nebraska Press, 2005).

410. Røde Kors Delegationens Besøg (Red Cross Delegation Visit), www.danskejoederitheresienstadt.org.

411. Dr. Eigil Henningsen, Beretning om Besøg i Theresienstadt, Fredag den 22 Juni 1944, Dansk Røde Kors (Report on Visit to Theresienstadt, Friday 22 June 1944, Danish Red Cross), Report dated 16 August 1944, www.danskejoederitheresienstadt.org.

412. S. Farré and Y. Schubert, L'Illusion de L'Objectif. Le Délégué du CICR Maurice Rossel et les Photographies de Theresienstadt (The illusion of the lens. ICRC delegate Maurice Rossel and the Theresienstadt photographs). *Le Mouvement Social*, **227**, 65 (2009).

413. Excerpt from the Report of Maurice Rossel of the MKČK about the visit to Terezín in 1944, www.collections.jewishmuseum.cz/index.php/Detail/Object/Show/object_id/134242.

414. S. Farré, *The ICRC and the Detainees in Nazi Concentration Camps (1942–1945)* (Cambridge University Press, 2013).

415. Nobel Prize for Peace for 1944, International Committee of the Red Cross, www.nobelprize.org/prizes/peace/1944/red-cross/facts/.
416. K. Margry, 'Theresienstadt' (1944–1945): The Nazi propaganda film depicting the concentration camp as paradise. *Hist. J. Film, Radio Television*, **12**, 145 (1992).
417. N. Drubek, The three screenings of a secret documentary. Theresienstadt Revised, in *Apparatus. Film, Media and Digital Cultures of Central and Eastern Europe*, no. 2–3 (2016).
418. K.-H. Schoeps, *Literature and Film in the Third Reich* (Camden House, Rochester, 2003).
419. The Trial of Karl Rahm, the Former Commandant of the Terezín Ghetto, Jewish Museum of Czechoslovakia, www.collections.jewishmuseum.cz/index.php/Detail/Object/Show/object_id/136398.
420. Salz, Professor Arthur (1881–1963), File 1933-47, MS. S.P.S.L. 238/2.
421. V. Ö. Vilhjálmsson and B. Blüdnikow, Rescue, expulsion and collaboration: Denmark's difficulties with its World War II past, *Jewish. Pol. Stud. Rev.*, **18**, 3 (2006).
422. Claude Lanzmann Shoah Collection, *Interview with Maurice Rossel*, United States Holocaust Memorial Museum, April-May 1979.
423. H. Arendt, *Eichmann in Jerusalem: A Report on the Banality of Evil* (Viking Press, New York, 1963).

Chapter 10: Conclusions

424. Peierls, Sir Rudolf Ernst (1907–95), File 1938-48, MS. S.P.S.L. 335/9 and 49/10.
425. D. Zimmerman, *Ensnared Between Hitler and Stalin. Refugee Scientists in the USSR* (University of Toronto Press, 2023).
426. S. Dorril, *Blackshirt: Sir Oswald Mosley and British Fascism* (Thistle Publishing, London, 2017).

Bibliography

E. Amaldi, *The Adventurous Life of Friedrich Georg Houtermans, Physicist (1903–1966)*, in S. Braccini, A. Eredidato and P. Scampoli (Eds.) (Springer, Heidelberg, 2012).

H. Arendt, *Eichmann in Jerusalem: A Report on the Banality of Evil* (Viking Press, New York, 1963).

P. Ball, *Serving the Reich: The Struggle for the Soul of Physics Under Hitler* (University of Chicago Press, 2014).

N. Bentwich, *The Rescue and Achievement of Refugee Scholars: The Story of Displaced Scholars and Scientists 1933–1952* (M. Nijhoff, The Hague, 1953).

W. Beveridge, *A Defence of Free Learning* (Oxford University Press, 1959).

A. Bhattacharya, *The Man From the Future: The Visionary Life of John von Neumann* (Allen Lane, London, 2021).

J. Borkin, *The Crime and Punishment of I.G. Farben* (Barnes and Noble, New York, 1978).

M. Born, *My Life: Recollections of a Nobel Laureate* (Taylor and Francis, London, 1978).

M. Born and A. Einstein, *The Born-Einstein Letters 1916–1955* (Macmillan Press, New York, 1971).

L. W. B. Brockliss, *The University of Oxford: A History* (Oxford University Press, 2016).

L. W. B. Brockliss (Ed.) *Magdalen College Oxford: A History* (Magdalen College, Oxford, 2008).

B. R. Brown, *Planck: Driven by Vision, Broken by War* (Oxford University Press, 2015).

D. C. Cassidy, *Uncertainty: The Life and Science of Werner Heisenberg* (W. H. Freeman, New York, 1992).

D. C. Cassidy, *Beyond Uncertainty: Heisenberg, Quantum Physics, and the Bomb* (Bellevue Literary Press, New York, 2010).

R. W. Clark, *Tizard* (MIT Press, Massachusetts, 1965).

D. C. Clary, *Schrödinger in Oxford* (World Scientific Publishing, Singapore, 2002).

F. Close, *Trinity: The Treachery and Pursuit of the Most Dangerous Spy in History* (Penguin, London, 2019).

B. Collins, *Robert Eisler and the Magic of the Combinatory Mind: The Forgotten Life of a 20th Century Austrian Polymath* (Palgrave Macmillan, New York, 2021).

R. M. Cooper (Ed.) *Refugee Scholars. Conversations with Tess Simpson* (Moorland Books, Leeds, 1992).

A. Dagan, G. Hirschler and L. Weiner (Eds.) *The Jews of Czechoslovakia. Historical Studies and Surveys*, Vol. III (Jewish Publication Society of America and the Society for the History of Czechoslovak Jews, Philadelphia, 1984).

J. M. Deem, *The Prisoners of Breendonk. Personal Histories from a World War II Concentration Camp* (Houghton Mifflin Harcourt, Boston, 2015).

M. Eckert, *Arnold Sommerfeld: Science, Life and Turbulent Times 1868–1951*, T. Artin, Trans. (Springer, Berlin/Heidelberg, 2013).

C. P. Enz, *No Time to Be Brief: A Scientific Biography of Wolfgang Pauli* (Oxford University Press, 2002).

B. Falk, *Caught in a Snare: Hitler's Refugee Academics, 1933–1949* (Melbourne University, 1998).

G. Farmelo, *The Strangest Man: The Hidden Life of Paul Dirac, Quantum Genius* (Faber and Faber, London, 2009).

J. Feichtinger, H. Matis, S. Sienell and H. Uhl, *The Academy of Sciences in Vienna 1938 to 1945* (Austrian Academy of Sciences Press, Vienna, 2014).

G. Ferry, *Max Perutz and the Secret of Life* (Chatto and Windus, London, 2007).

R. Fox and G. Gooday (Eds.) *Physics in Oxford 1839–1939* (Oxford University Press, 2005).

O. Frisch, *What Little I Remember* (Cambridge University Press, 1979).

K. Gavroglu, *Fritz London: A Scientific Biography* (Cambridge University Press, 1995).

P. Gillman and L. Gillman, *Collar the Lot! How Britain Interned and Expelled Its Wartime Refugees* (Quartet Books, London, 1980).

A. Gottwaldt and D. Schulle, *Die 'Judendeportationen' aus dem Deutschen Reich 1941–1945* (Marix Verlag, Wiesbaden, 2005).

N. T. Greenspan, *The End of the Certain World: The Life and Science of Max Born* (Basic Books, New York, 2005).

N. T. Greenspan, *Atomic Spy: The Dark Lives of Klaus Fuchs* (Viking, New York City, 2020).

A. Grenville, *Refugees from the Third Reich in Britain* (Rodopi, New York, 2002).

J. Gribbin, *Erwin Schrödinger and the Quantum Revolution* (Transworld, London, 2012).

E. Y. Hartshorne, *The German Universities and National Socialism* (George Allen & Unwin, London, 1937).

J. L. Heilbron, *The Dilemmas of an Upright Man: Max Planck and the Fortunes of German Science* (Harvard University Press, Massachusetts, 2000).

R. Highfield and P. Carter, *The Private Lives of Albert Einstein* (Faber and Faber, London, 1993).

C. Hoff, *Anna und Leon* (Hentrich & Hentrich, Berlin, 2005).

D. Hoffmann and M. Walker (Eds.) *The German Physical Society in the Third Reich: Physicists between Autonomy and Accommodation*, A.M. Hentschel, Trans. (Cambridge University Press, 2007).

W. Höflechner, *History of the Karl-Franzens University Graz: From the Beginning until 2005* (Leykam, Graz, 2006).

M. A. Hutton, *Testimony from the Nazi Camps. French Women Voices* (Routledge, London, 2004).

J. James, T. Steinhauser, D. Hoffmann and B. Friedrich, *One Hundred Years at the Intersection of Chemistry and Physics. The Fritz Haber Institute of the Max Planck Society 1911–2011* (De Gruyter, Berlin, 2011).

L. P. Johannsen, *Erich Aschenheim, Albert Eckstein, Julius Weyl. Jüdische Pädiater Im Vorstand Der Vereinigung Rheinisch Westfälischer Kinderärzte* (Hentrich & Hentrich, Berlin, 2010).

E. Jones, *The Life and Work of Sigmund Freud* (Pelican Books, London, 1964).

Y. Kapp and M. Mynatt, *British Policy and the Refugees, 1933–1941* (Routledge, London, 1997).

G. Kerber, A. Dick and W. Kerber, *Erwin Schrödinger 1887–1961: Documents, Materials and Pictures* (Austrian Central Library for Physics, Vienna, 2015).

C. W. Kilmister (Ed.) *Schrödinger: Centenary Celebrations of a Polymath* (Cambridge University Press, 1987).

S. Klarsfeld, *Le Calendrier de la Persécution des Juifs de France. Septembre 1942 — Août 1944. La Shoah en France* (Fayard, Paris, 2001).

S. Klarsfeld, *Die Endlösung der Judenfrage in Frankreich. Deutsche Dokumente 1941–1944* (Klarsfeld, Paris, 1977).

A. Kramish, *The Griffin: The Greatest Untold Espionage Story of World War II* (Houghton, Boston, 1986).

S. Lee, *Sir Rudolf Peierls: Selected Private and Scientific Correspondence*, Vols. 1 and 2 (World Scientific, Singapore, 2007).

L. Leff, *Well Worth Saving. American Universities' Life-and-Death Decisions on Refugees from Nazi Europe* (Yale University Press, New Haven, 2019).

M. A. Livingston, *The Fascists and the Jews of Italy. Mussolini's Race Laws, 1938–1943* (Cambridge University Press, 2014).

L. London, *Whitehall and the Jews, 1933–1948: British Immigration Policy, Jewish Refugees and the Holocaust* (Cambridge University Press, 2000).

S. Marks, P. Weindling and L. Wintour (Eds.) *In Defence of Learning: the Plight, Persecution and Placement of Academic Refugees, 1933–1980s* (Oxford University Press for the British Academy, 2011).

K. D. McRae, *Nuclear Dawn: F. E. Simon and the Race for Atomic Weapons in World War II* (Oxford University Press, 2014).

J. Medawar and D. Pyke, *Hitler's Gift: The True Story of the Scientists Expelled by the Nazi Regime* (Arcade, New York, 2000).

K. Mendelssohn, *The World of Walther Nernst: The Rise and Fall of German Science, 1864–1941* (University of Pittsburgh Press, 1973).

K. von Meyenn (Ed.) *Eine Entdeckung von ganz ausserordentlicher Tragweite. Schrödingers Briefwechsel zur Wellenmechanik und zum Katzenparadoxon*, Band 1 and 2 (Springer, Berlin/Heidelberg, 2011).

W. J. Moore, *Schrödinger: Life and Thought* (Cambridge University Press, 1989).

J. Morrell, *Science at Oxford 1914–1939: Transforming an Arts University* (Oxford University Press, 1997).

P. Neville, *Hitler and Appeasement: The British Attempt to Prevent the Second World War* (Hambledon, London, 2006).

A. Pais, *Subtle Is the Lord: The Science and the Life of Albert Einstein* (Oxford University Press, 2005).

K. Popper, *Unended Quest: An Intellectual Autobiography* (William Collins, Glasgow, 1974).

A. Roberts, *The Storm of War: A New History of the Second World War* (Allen Lane, London, 2009).

A. Robinson. *Einstein on the Run: How Britain Saved the World's Greatest Scientist* (Yale University Press, New Haven, 2019).

L. Rothkirchen, *The Jews of Bohemia and Moravia: Facing the Holocaust* (University of Nebraska Press, 2005).

R. Rurup and M. Schüring, *Schicksale und Karrieren. Gedenkbuch für die von den Nationalsozialisten aus der Kaiser-Wilhelm-Gesellschaft vertriebenen Forscherinnen und Forscher. Series: History of the Kaiser Wilhelm Society under National Socialism*, Vol. 14 (Wallstein Verlag, Göttingen, 2008).

G. Schneider, *Exile and Destruction. The Fate of Austrian Jews 1938–1945* (Praeger, Westport CT, 1995).

K.-H. Schoeps, *Literature and Film in the Third Reich* (Camden House, Rochester, 2003).

E. Schrödinger, *What Is Life? The Physical Aspect of the Living Cell* (Cambridge University Press, 1944).

D. N. Schwartz, *The Last Man Who Knew Everything, the Life and Times of Enrico Fermi, Father of the Nuclear Age* (Basic Books, New York, 2017).

W. T. Scott, *Erwin Schrödinger: An Introduction to his Writings* (University of Massachusetts Press, 1967).

J. Seabrook, *The Refuge and the Fortress: Britain and the Flight from Tyranny* (Palgrave Macmillan, London, 2008).

M. Shifman, *Physics in a Mad World* (Based on the writings of V. J. Frenkel) (World Scientific Publishing, Singapore, 2015).

R. Siegmund-Schultze, *Mathematicians Fleeing from Nazi Germany, Individual Fates and Global Impact* (Princeton University Press, 2009).

R. L. Sime, *Lise Meitner: A Life in Physics* (University of California Press, Berkeley, 1996).

D. Snowman, *The Hitler Emigrés. The Cultural Impact on Britain of Refugees from Nazism* (Chatto & Windus, London, 2002).

A. Speer, *Inside the Third Reich* (Weidenfeld & Nicolson, London, 1970).

M. Szöllösi-Janze (Ed.) *Science in the Third Reich* (Berg, Oxford, 2001).

T. Tchernavin, *Escape from the Soviets* (Tr. N. Alexander) (E.P. Dutton, New York, 1934).

V. V. Tchernavin, *I Speak for the Silent Prisoners of the Soviets* (Tr. N.M. Oushakoff) (Hale, Cushman and Flint, New York, 1935).

J. B. Tucker, *War of Nerves. Chemical Warfare from World War I to Al-Qaeda* (Pantheon Books, New York, 2006).

L. Vastenhout, *Between Community and Collaboration: 'Jewish Councils' in Western Europe under Nazi Occupation* (Cambridge University Press, 2022).

A. Vogt, *Vom Hintereingang zum Hauptportal? Lise Meitner und ihre Kolleginnen an der Berliner Universität und in der Kaiser-Wilhelm-Gesellschaft* (Franz Steiner Verlag, Stuttgart, 2007).

M. Walker, *German National Socialism and the Quest for Nuclear Power, 1939–49* (Cambridge University Press, 1989).

V. Weisskopf, *The Joy of Insight: Passions of a Physicist* (Basic Books, New York, 1991).

E. P. Wigner, *The Recollections of Eugene P. Wigner, as told to Andrew Szanton*, (Springer, New York, 1992).

R. E. Wolman, *Crossing over. An Oral History of Refugees from Hitler's Reich* (Twayne Publishers, New York, 1996).

D. Zimmerman, *Ensnared Between Hitler and Stalin. Refugee Scientists in the USSR* (University of Toronto Press, 2023).

Figures and Permissions

2.1 Friedrich Duschinsky in his Belgian Aliens Registration Certificate of 1934. Foreigners' file No. A106.433. © National Archives of Belgium.

2.2 Letter from Duschinsky congratulating his teacher Schrödinger on his Nobel Prize, written from Gablonz on 11 November 1933. *Source*: Austrian Central Library for Physics. Reprinted with kind permission of the Braunizer family.

2.3 Peter and Emilia Pringsheim in their Belgian Aliens Registration Certificates of 1933. Foreigners' file No. 1.146.765. © National Archives of Belgium.

2.4 Siegwart Hermann, inventor of Kombucha (~1914).

2.5 List of prisoners in Drancy on staircase 2 who were transported to Auschwitz on 11 November 1942. Note that "Buchinsky" should be "Duschinsky". *Source*: C45_21: p. 21 du convoi 45 parti du camp de Drancy le 11/11/1942 à destination du camp d'Auschwitz-Birkenau. Permission to publish granted by Mémorial de la Shoah, Paris (France).

2.6 Memorial in Bagneaux-sur-Loing naming those arrested and deported by the Nazis. The names include Bedrich Buchinsky and Alexandre, Hilda and Maximilien Altmann. © Maryvonne Braunschweig.

2.7 Emanuel Wasser and his wife Sara Zylberszac in their Belgian Aliens Registration Certificates of 1939. Foreigners' file No. A326.560. © National Archives of Belgium.

2.8   Elisabeth Wasser, photographed on 29 April 1944 by the Police Department in Switzerland. File N21992 © Swiss Federal Archives, Federal Department of Home Affairs, Berne.

2.9   Memorial at the University of Strasbourg listing those from the University who were deported, shot or murdered in World War II. "Emmanuel Wasser" is listed. Permission to publish granted under the GNU Free Documentation Licence.

2.10   Group at the Kaiser-Wilhelm-Institut für Physikalische Chemie und Elektrochemie, Berlin-Dahlem, in the 1920s. Marie Wreschner is in the second row, second from the right. © Max Planck Society, Berlin.

2.11   Patent published by Marie Wreschner and Laurence Loeb in 1930 describing a new treatment for cancer.

2.12   Alfred Byk (~1910). With kind permission of Professor Rosemary Pattenden, granddaughter.

2.13   Alfred Byk (~1938). With kind permission of Professor Rosemary Pattenden, granddaughter.

2.14   Erich Lehmann (from an article by J. Eggert).

3.1   Paul and Irène Dreyfuss in their Belgian Aliens Registration Certificates, 1937. Foreigners' file No. A271.228. © National Archives of Belgium.

3.2   Paul Dreyfuss, second row on the left, in the St. Cyprien internment camp in the hot summer of 1940 (upper photograph). The lower photograph shows the crowded accommodation in the camp. Courtesy of Judy Navon-Dreyfuss and Michael Dreyfuss, daughter and son.

3.3   Walter Michaelis in his Belgian Aliens Registration Certificate. Foreigners' file No. A159.669. © National Archives of Belgium.

3.4   Dr. Bela Gaspar in Laboratory, Hollywood (~1944) (top photograph). Photograph by Richard C. Miller. Gasparcolor print. Gift of the Miller Family Trust © Miller Family Trust A.
Dr. Paul Dreyfuss in Laboratory, Hollywood, undated (bottom photograph). Photograph by Richard C. Miller. Courtesy of Judy Navon-Dreyfuss and Michael Dreyfuss, daughter and son.

3.5 Vladimir Lasareff in his Belgian Aliens Registration Certificate, 1933. Foreigners' file No. A278.332. © National Archives of Belgium.

3.6 Paul and Kathe Goldfinger (née Deppner) in their Belgian Aliens Registration Certificates of 28 October 1939, subsequently stamped with "Juif-Jood" in his case. Foreigners' file No. A87.144. © National Archives of Belgium.

3.7 Boris and Lidja (née Gabis) Rosen in their Belgian Aliens Registration Certificates of 26 September 1935. Foreigners' file No. A84.104. © National Archives of Belgium.

3.8 Alfred Lustig. With kind permission of Professor Mansur Gilmullin.

3.9 Alfred Lustig with his wife Lydia and children Mikhail and Zinaida in Yelabuga, Russia. With kind permission of Professor Mansur Gilmullin.

3.10 Giulio Bemporad in 1943.

3.11 Karl-Heinrich Riewe, photograph used in his AAC application (1935). Oxford, Bodleian Libraries MS S.P.S.L 337/8. © Council for At-Risk Academics.

3.12 Fritz Houtermans in a Soviet prison in 1938.

3.13 Fritz and Charlotte Houtermans in 1932. With kind permission from the University of Berne.

4.1 Los Alamos badge photograph of Otto Frisch, 1943. Unless otherwise indicated, this information has been authored by an employee or employees of the Los Alamos National Security, LLC (LANS), operator of the Los Alamos National Laboratory under Contract No. DE-AC52-06NA25396 with the US Department of Energy. The US Government has rights to use, reproduce, and distribute this information. The public may copy and use this information without charge, provided that this Notice and any statement of authorship are reproduced on all copies. Neither the Government nor LANS makes any warranty, express or implied, or assumes any liability or responsibility for the use of this information.

4.2   Klaus Fuchs after his arrest by the British Security Services in 1950.

5.1   Ludwig and Hedwig Berwald (née Adler), 1922. With kind permission of Národní archiv, Policejní ředitelství Praha II — všeobecná spisovna (National Archives, Police Headquarters Prague II. — General Register), the period 1931–1940, record signature B 1564/2, box 4698, Ludwig Berwald, born 8.12.1883; and record signature B 1564/1, box 4698, Hedwig Berwald, born 12.9.1875.

5.2   Walter and Elise Fröhlich (née Goliath), 1927. With kind permission of Národní archiv, Policejní ředitelství Praha II — všeobecná spisovna (National Archives, Police Headquarters Prague II. — General Register), the period 1941–1950: record signature F 1598/3, box 2330, Walter Fröhlich, born 2.12.1902; and record signature F 1606/5, box 2332, Elise Fröhlich, born 4.5.1907.

5.3   Robert Remak (~1933).

5.4   Otto Blumenthal as a student in Göttingen (left) and a professor at Aachen (right).

6.1   Arthur Simons, stills from his film *Head Positions and Muscle Tone* (1923). © Springer Nature and Copyright Clearance Center.

6.2   Otto Sittig in his SPSL application of 1938. Oxford, Bodleian Libraries MS S.P.S.L 552/2. © Council for At-Risk Academics.

6.3   Ferdinand Blumenthal in 1930.

6.4   Hans Hirschfeld (~1938). © Ullstein Bild.

6.5   Hans Hirschfeld in Theresienstadt, 4 August 1943. Sketch by Max Plaček (1902–1944). Pencil and colour pencil on paper. Collection of the Yad Vashem Art Museum, Jerusalem. Gift of Hermann Weiss, courtesy of Stephen Barber, Canada. Photo © Yad Vashem Art Museum.

6.6   (a) Front page of *Handbuch der allgemeinen Hämatologie*, edited by Hans Hirschfeld and Anton Hittmair, 1932 (left).

(b) Front page of *Handbuch der gesamten Hämatologie*, edited by Ludwig Heilmeyer and Anton Hittmair, 1957 (right).

7.1   Hellmuth Simons, March 1943. File Ar 126.5 © Schweizerisches Sozialarchiv.

7.2   Hellmuth Simons with Alis Guggenheim in Switzerland, 1943. F 5090-Fa-101 Guggenheim, Alis (1896–1958) Schweizerisches Sozialarchiv. © Courtesy estate Olivia Heussler, Zürich.

7.3   Hellmuth Simons, 1945. File Ar 126.5 © Schweizerisches Sozialarchiv.

7.4   Vladimir and Tatiana Tchernavin, 1933.

8.1   Alfred Rheinheimer in his AAC application (1933). Oxford, Bodleian Libraries MS S.P.S.L 245/8. © Council for At-Risk Academics.

9.1   Robert Eisler (1928), in the Abraham Schwadron Collection at the National Library of Israel.

9.2   Paul Eppstein: *Reichsvertretung der Juden in Deutschland (1936–1938)*. © Leo Baeck Institute, F 39694.

To the knowledge of the author, all photographs are in the public domain unless stated otherwise.

Permissions for
Letters and Quotes

The author acknowledges the kind permission and assistance to publish photographs, letters and quotes:

Alfred Byk (Professor Rosemary Pattenden, granddaughter); Paul Dreyfuss (Michael Dreyfuss and Judy Navon-Dreyfuss, son and daughter); Alfred Lustig (Mansur Gilmullin, colleague of Alfred Lustig); Marie Wreschner (Archives of the Max Planck Society, Berlin, Susanne Uebele); Academic Assistance Council and the Society for Protection of Science and Learning (the Council for At-Risk Academics, Stephen Wordsworth, Executive Director); National Archives of Belgium (Filip Strubbe); Mémorial de la Shoah (Dorotheé Boichard); Memorial at Bagneaux-sur-Loing (Cercle d'étude de la Déportation et de la Shoah, Nicole Mullier and Maryvonne Braunschweig); Swiss Federal Archives, Federal Department of Home Affairs, Berne; Schweizerisches Sozialarchiv and Archive of Alis Guggenheim (Olivia Heussler, Zürich); the University of Berne (Niklaus Bütikofer, Universitätsarchiv); Austrian Central Library for Physics, Vienna, and the Braunizer family; the Richard C. Miller Family Trust (Jacklyn Burns, J. Paul Getty Museum); the National Archives of Czechoslovakia in Prague (Kateřina Vansová); Ullstein Bild (Topfoto, Ioan Nedelcu); United States Holocaust Memorial Museum; Yad Vashem Art Museum; Leo Baeck Institute (Mareike Hennies); Amsterdam City Archives; Churchill Archives Centre, Churchill College, Cambridge; UK National Archives, Kew; Special Collections and Archives, University at

Albany (Melissa McMullen); and Records of the Emergency Committee in Aid of Displaced Foreign Scholars, The Brooke Russell Astor Reading Room for Rare Books and Manuscripts, New York Public Library, Astor, Lenox, and Tilden Foundations (Cara Dellatte, Reference Archivist).

Reproductions of reports in *Nature*, the *Annalen der Physik* and *Naturwissenschaften* with permission through the Copyright Clearance Center RightsLink Service.

Dates of birth and curriculum vitae details of the scientific refugees were obtained from the SPSL archives. Stated dates of death were obtained from a variety of sources as referenced.

Index

A

Abderhalden, Emil, 48
Adams, Walter, 2, 37, 105, 168, 170, 226
Altmann, Salomon, 232
American Association of University Women, 56
Anderson, Lindsay, 227
Annals of Mathematics, 132
Anschluss, 4, 34–35, 38, 96, 114, 167, 223–224, 230
Aschenheim, Erich
    AAC application, 161
    Aschenheim, Annemarie, 161, 162
    Aschenheim, Anni, 163
    Aschenheim, Emma, 161
    Aschenheim, Eva, 164
    Aschenheim, Gabriele, 163
    birth, 160
    Carr-Saunders, 162
    Fraser, 161
    Gestapo arrest, 162
    Krailling, 161
    position in Dusseldorf, 160
    Schlossmann, 160
    Semon, 160–162
    suicide, 162
    work in Remscheid, 161
Auschwitz, 24–27, 39–40, 65, 68, 77, 85, 137, 146, 156, 160, 170, 180, 218, 233, 235–236, 242

B

Bagneaux-sur-Loing, 22–23, 26, 29
Balliol College, Oxford, 158, 220, 227
Barraquer–Simons syndrome, 155
Basel, 80, 188, 194–195, 197
Beer, Arthur, 135
Behrend, Felix, 65
Beiglböck, Wilhelm, 178
Belgrade, 166–170
Bemporad, Giulio
    anti-Semitism, 101
    Bemporad, Azeglio, 100
    birth, 99
    Carloforte, 100
    death, 103
    Delasem, 102
    in Rome, 103
    Mussolini, 100–103
    Naples Observatory, 100

Simpson letters, 102
SPSL application, 101
Turin Observatory, 100
University of Catania, 99
Volta, 102–103
Beria, Laurentia, 112, 121
Bernal, J. D., 111
Berne, 106, 120, 194–195
Berwald, Ludwig
    Berwald, Hedwig, 130, 132
    birth, 130
    contributions to mathematics,
        132
    dismissed, 131
    Finsler and Cartan Geometry,
        132
    Fröhlich, 133
    German University in Prague,
        130
    Lodz Ghetto, 132
    Ludwig Maximilian University,
        130
    Simpson letters, 132
    SPSL application, 131
    Whitehead letter, 131
Bethe, Hans, 5, 107
Beveridge, Lord, 2, 226
biological warfare, 186, 196
biologists, 181
Birnbaum, Suzanne, 40
Blackett, Patrick, 33, 108–111, 116,
    124
Blumenthal, Ferdinand
    Adams response, 169
    Belgrade move, 166–170
    birth, 164
    Blumenthal, Herma, 168, 170
    Charité Hospital, Berlin, 164

Cramer, 168, 170, 172
Estonia move, 170
Fränkel, 170
Freiburg doctorate, 164
Friedrich Wilhelms University,
    164
Hirschfeld, 165
leader in cancer research, 165
Narvas death, 170
Simpson letters, 170
SPSL concerns, 167
Vienna move, 167
Blumenthal, Otto
    AAC application, 147
    birth, 147
    Bishop of Chichester, 148
    Blumenthal, Ernst, 147
    Blumenthal, Mali, 147, 149
    Blumenthal, Margrete, 148
    Born, 150
    death, 149
    Emergency Committee, 147
    Freudenthal, 148
    Hardy letter, 147
    Hilbert student, 147
    Magdalen College, Oxford, 148
    mathematical contributions, 149
    Mathematische Annalen Editor,
        147
    remembered in Aachen, 149
    return to Netherlands, 148
    Simpson letters, 148
    Theresienstadt, 149
    University of Aachen, 147
    visit to England, 148
    Westerbork camp, 149
    Whitehead letter, 148
Bohr, Harald, 110

Bohr, Niels, 112, 124–125
Bond, James, 92
Born, Max
  Blumenthal, Otto, 150
  Byk comments, 59, 61
  Cambridge move, 3
  Duschinsky comment, 16
  Edinburgh chair, 36
  Fuchs, 128
  Houtermans, 116
  left Germany, 1
  Newton-John family, 117
  relative victims, 53
  Wasser, 32
Bosch, Carl, 4
Bragg, William H., 2
Breendonk, 77, 84–87
Breslau, 55, 66–68, 109, 158, 189,
  220
British Federation University
  Women, 6, 50, 54
British Medical Association, 153,
  240–241
Brunner, Alois, 39
Buchenwald, 85, 221, 225, 228–229
Bulloch, William, 176
Burkhardt, Heinrich, 130
Byk, Alfred
  AAC application, 61
  Anna Graeffner suicide, 63
  Behrend, 65–66
  birth, 57
  Born comments, 59, 61
  Byk, Hedwig, 58, 63
  Byk, Heinrich, 57
  Byk, Hilde, 58, 62–63, 66
  Byk, Marianne, 58, 62–66
  Byk, Suse, 57

  Declaration of professors, 58
  education, 58
  Einstein, 59
  Einstein letter to daughters, 63
  Emergency Committee, 59
  Eppstein, 60
  Fischer, 58
  Foley, 62
  General Electric Company, 59
  Handbook of Physics, 59
  Hedwig Byk suicide, 63
  Murrow, 60
  Nernst, 58, 61
  Olivaer Platz commemoration,
    66
  Pattenden, Rosemary, 66
  Planck, 58, 61
  Prussian Professor, 58
  Red Cross message, 64
  Rosenberg message, 64
  Sobibór murder, 65
  SPSL application, 62
  University of Berlin, 58
  Ursell enquiries, 65
  Weizmann, 61
  Windaus, 64

C
Cairns, Hugh, 155
Camp-des-Milles, 38, 77, 192
Carmichael, E. Arnold, 158
Carr-Saunders, Alexander, 162
Cassirer, Ernst, 230
Chain, Ernst, 153
Chaplin, Charlie, 207
Charité Hospital, Berlin, 155, 164,
  172, 175
Chemical Physics Letters, 91

chemical weapons, 186
chemists, 9, 181
Chromogen, 78–80
Churchill, Winston, 5, 125, 193
CIA, 120
Cibachrome, 80
Clark, Kenneth, 224, 230
Clermont-Ferrand, 39, 41–42, 194
Cohn, Benno, 237
Cohn, Ernst J., 220
Courant, Richard, 138
Courtauld Institute of Art, 210
Cramer, William, 168, 170, 172
Croydon airport, 20
Curie, Marie, 11
Czech Refugee Trust Fund, 27, 135

D
Dachau, 221, 224, 227–229
Dale, Henry, 35
D-Day, 196
De Gaulle, Charles, 193
Demuth, Fritz, 17
Dingle, Herbert, 56
Dirac, Paul, 34, 42
Donnan, Frederick, 14, 47, 49
Drancy, 23–26, 39–40, 44, 192, 194
Dreyfuss, Paul
    AAC application, 74
    Belgium, 75
    birth, 73
    Brussels arrest, 75
    Camp-des-Milles, 77
    Chromogen, 78–80
    Cibachrome, 80
    Dreyfuss, Irene, 74, 77–78
    Dreyfuss, Judy, 77
    Dreyfuss, Laura, 74–75, 77, 85

Dreyfuss, Maurice, 73
Dreyfuss, Michael, 77
    Emergency Committee, 74
    emigration to USA, 77
    Gaspar, 75, 77–80
    Goldfinger, 78
    Michaelis family, 77
    Miller, 78
    Pfeiffer, 74
    Pringsheim, 73, 75
    SPSL, 80
    St. Cyprien, 75
    University of Bonn, 74
    University of Cagliari, 74
    University of Catania, 74
Dulles, Allen, 195
Duschinsky, Erich, 19–23, 26–28
Duschinsky, Fritz
    AAC application, 14
    Altmann, Alexandre, 20, 26, 29
    Altmann company, 19
    Altmann, Hilda, 19–21, 26, 29
    Altmann, Maximilian, 19–20,
      26, 29
    Auschwitz, 25–26
    Bagneaux-sur-Loing, 22–23, 29
    birth, 10
    Born comment, 16
    Brussels, 14
    Buchinsky, Bedrich, 23–25, 29
    Croydon airport, 20
    dismissed from Berlin, 13
    Drancy, 23–26
    Duschinsky, Erich, 19–23,
      26–28
    Duschinsky, Jenny, 11, 26, 233
    Duschinsky rotation, 18, 30
    education, 11

Eliaschewitsch, 18
Emergency Committee, 14, 23
fluorometry theory, 11, 30
Gablonz, 10–14, 19–27, 44
Gestapo arrest, 23
Hermann, 21–23
Leningrad, 17–20, 27, 30
Nernst reference, 15–16
Paris, 20
patents, 19, 21
Piccard, 13, 14
Pringsheim, 11, 13–18, 20–21, 28–29
Schrödinger, 11–12, 14, 18
Shoah Memorial, 29
University of Berlin, 11
Ursell enquiries, 27–29
Weissenberg, 12–15, 27
Zeitschrift für Physik papers, 11
Düsseldorf, 155, 160, 162, 182–183, 185, 192, 196, 198–199

E
Eastman Kodak, 37, 56, 71
Eckhart, Ludwig, 151
Edinburgh, 36, 116, 135–136, 162, 211
Ehrenhaft, Felix, 31–33, 36–37, 42, 44, 95–96
Eichmann, Adolf, 25, 39, 97, 132, 156, 233, 237, 242
Einstein, Albert
Berlin, 11
Byk daughters letter, 63
Byk paper, 57, 59
Byk support, 59
Ehrenhaft, 32, 36
Eisler, 222–224

Ettlinger correspondence, 83, 85
Lasareff, 83, 85–86
pacifism signature, 58
Prague, 134–135
Pribram letter, 223
Pringsheim, 11
reference letters, 3
Schrödinger and Eisler, 230
Simons reference, 183
Simpson letters, 85
Soviet Ambassador letter, 111
Suse Byk portraits, 58
The Times letter, 209
USA, 5
Wasser letter, 32
Eisenhower, General, 196
Eisler, Robert
arrested in Udine, 222
Art history, 222
birth, 221
books on money, 223
Dachau and Buchenwald, 221, 224, 229
death, 229
Einstein letters, 222, 224
Eisler, Friedrich, 221
Eisler, Lili, 224–229
Eisler, Otto, 227
Eisler, Rosalia, 222
Evans, 225, 228
five essays, 222
Gordon, 225
Hitler's name, 228
interned, 225
League of Nations position, 222
letters to *The Times*, 227–229
Lindsay Anderson diary, 227

Man into Wolf, 228
Murray, 222–224, 228, 230
obituaries, 229
Schrödinger similarities,
   229–230
Simpson letters, 226
SPSL application, 223
The Royal Art of Astrology, 227
University of Vienna, 222
Ursell letters, 228
visit to Oxford in 1922, 222
Wilde Readership Oxford, 224–
   225, 228
Elektrostal Laboratory, 121
Eliaschewitsch, Michael, 18
Ellis Island, 77, 199
Emergency Committee New York,
   5–6, 14, 17, 23, 52, 59–61, 70,
   74, 135, 147, 188
engineers, 215
Eppstein, Paul
   AAC application, 232
   birth, 231
   Byk, 60
   Chair Theresienstadt Jewish
      Council, 233
   Cohn statement, 237
   death, 235
   Eichmann, 233, 237, 242
   Eppstein, Hedwig, 236, 242
   Lederer, 232–233
   Mannheim, 231–232
   Rahm, 233, 236
   Red Cross visit, 233–237
   Reich Representative German
      Jews, 232
   Salz, 236
   Simpson letters, 236

Theresienstadt, 233–236, 242
   University of Heidelberg, 231
Erdélyi, Arthur, 136
ETH, 57, 87
Ettlinger, Lionel, 83, 85–86
Evans, Sir Arthur, 225, 228

F
Fano, Gino, 101
Fano, Ugo, 101
Faraday Discussion, 17
FBI, 200
Fermi, Enrico, 16, 101, 107, 115
Feynman, Richard, 110
Fischer, Emil, 58
Fleming, Ian, 92
Flexner, Abraham, 14
Flynn, John, 192
Fock, Vladimir, 18, 109, 118
Foges, Wolfgang, 167
Foley, Frank, 62
Fondation Francqui, 6, 82, 175
Foreign Legion, 192, 194, 197, 201
Fowler, Ralph, 105
Fox, H. Munro, 207
Franck, James, 11, 18, 47, 55,
   89–90, 93–94, 103, 106–107,
   110, 179
Fränkel, Ernst, 165, 168–170, 176
Fraser, Sir Francis, 161
Free University of Brussels, 13–14,
   37, 91
Freiburg, 58, 74, 77, 137, 164, 178,
   182, 189
French Resistance, 193
Freudenthal, Hans, 142–144,
   148–149
Freud, Sigmund, 229–230

Freundlich, Herbert, 45, 47, 49–52, 55, 83, 87, 89–90, 134, 166
Frick Library, 211
Friedrich-Wilhelms University, 55, 164, 172
Frisch, Otto
  Houtermans, 108
  letter from Sante Fe, 123
  Oliphant letter, 125
  paper with Meitner, 124
  Professorship at Cambridge, 126
  report with Peierls, 124
Fröhlich, Cecilie, 54, 57
Fröhlich, Herbert, 33–34
Fröhlich, Walter
  Berwald, 133
  birth, 133
  death, 136
  dismissal, 134
  Emergency Committee, 135
  Fröhlich, Elise, 133, 135–136
  German University in Prague, 133
  Lodz Ghetto, 136
  MacRobert grant, 135
  Simpson letter, 134
  SPSL application, 134
  visa, 135
  Whitehead letter, 134
  yellow star, 136
Fuchs, Karl
  Birmingham appointment, 128
  Born letter, 128
  criticism of AAC/SPSL, 127
  Gold, 127
  interned, 125, 128
  letter from Sante Fe, 126
  Simpson letters, 127–128
  trial for espionage, 127
Funk, Paul, 134, 150, 233

G
Gablonz, 10–14, 19–27, 44
Gamow, George, 106
Gaspar, Bela, 75, 77–80, 89–91
Gecow, Leon, 41
Gerlach, Walter, 114, 116
German Physical Society, 66, 103, 122
German SS, 188, 193
German University in Prague, 11, 55, 130–131, 133, 150, 157
Gerron, Kurt, 235–236, 242
Gestapo, 23, 39, 44, 63, 114, 116–117, 119, 147, 156, 162, 167, 170, 233, 237
Gibson, Charles, 2
Glasgow, 116, 135, 215–217
Goldfinger film, 92
Goldfinger, Paul
  AAC application, 81, 89
  Belgium, 88
  birth, 87
  Chemical Physics Letters, 91
  ETH, 87
  Franck, 89–90
  Free University of Brussels, 91
  Freundlich, 87
  Gaspar, 89–90
  Goldfinger, Ernö, 89, 92
  Goldfinger, Georges, 88
  Goldfinger, Kate, 87, 91
  Goldfinger, Marianne, 88, 91
  Goldfinger, Oskar, 87
  Goldfinger, Peter, 91
  Goldfinger, Regine, 87

Haber, 87
James Bond film, 92
Kaiser-Wilhelm-Institut, 87
Kuhn, 87, 89
Lasareff, 88
Lasareff and Rosen, 80, 88
mass spectrometry, 91
papers, 91
patent, 89, 91
Pauli, 88, 92
Resistance, 91
Simpson letters, 90
University of Liège, 88–89
Ursell enquiries, 90
yellow star, 91
Yves Cape grandson, 91
Gordon, George, 35, 225
Göttingen, 1, 3, 16, 103, 107, 110,
   115–116, 120, 146–147, 150, 121
Grelling, Kurt, 151
Guggenheim, Alis, 196–197
Gurs, 28, 77
Guttmann, Ludwig, 158, 160

H
Haber, Fritz, 3, 5, 11, 45, 47, 54,
   58, 82, 87, 89, 92, 166
Hahn, Kurt, 141–142, 144
Hahn, Otto, 11, 119, 121
Halifax, Lord, 35
Hamburg, 137, 216, 220
Handbook of General Haematology,
   172, 178
Hardy, G. H., 3, 65, 129, 138–144,
   147
Hartogs, Fritz, 151
Hausdorff, Felix, 151
Head, Henry, 158

Heidelberg, 89, 160, 232, 236
Heilbronn, Hans, 129–130
Heilmeyer, Ludwig, 178–179
Heinrichsohn, Ernst, 25
Heisenberg, Werner, 16, 58,
   115–116, 119
Hellmann-Feynman theorem, 110
Hellmann, Hans, 110
Henderson, Nevile, 35
Henningsen, Eigil, 233, 236
Henri, Victor, 80–85, 89, 91, 93
Hermann, Siegwart, 21–23
Hermitage Museum, 203–204,
   210–211
Hertz, Gustav, 106, 121
Herxheimer, Herbert, 175–176
Hess, Victor, 35–36
Heydrich, Reinhard, 132, 136
Hilbert, David, 3, 138, 146–147,
   149–150
Hill, A. V., 172–173, 175
Himmler, Heinrich, 39, 156
Hinshelwood, Cyril, 51
Hirschfeld, Hans
   AAC application, 172
   birth, 172
   Blumenthal, 172
   Charité Hospital, Berlin, 172
   Cleghorn Thomson letter, 174
   Cramer, 172
   death, 178
   Friedrich-Wilhelms University,
      172
   Haematological Society, 172
   *Handbook of General
      Haematology*, 172
   Heilmeyer, 178–179
   Herxheimer, 175–176

Hill, 172, 175
Hirschfeld, Ilse, 173, 178
Hirschfeld, Kate, 174, 176, 178
Hirschfeld, Rosa, 172, 178
Hittmair, 172, 178–179
  property confiscation, 177
  Simpson letters, 173, 176
  SPSL application, 173
  Theresienstadt deportation, 176
  Ulm dedication, 179
Hitler, Adolf, 1–3, 5, 13, 19, 83, 96,
  99–100, 109, 132, 134, 138, 147,
  153, 155, 158, 165, 228, 243
Hittmair, Anton, 172, 178–179
Hoffmann La Roche, 188, 193–195
Hollitscher, Erna, 54
Houtermans, Charlotte, 107–114,
  120
Houtermans, Fritz
  AAC, 108
  Atkinson paper, 106
  Bernal, 111
  Bethe, 107
  birth, 106
  Blackett, 108
  Bohr, Harald letters, 110
  Born letter, 116
  death, 120
  Einstein letter, 111
  Fock, 109
  Franck, 106
  Frisch, 108
  Gamow paper, 106
  Gestapo prisoner swap, 114
  Göttingen position, 116
  Hellmann, 110
  Hertz, 106
  His Master's Voice, 108

Houtermans, Charlotte,
  107–114, 120
  Joliot-Curie letter to Stalin, 112
  Kapitsa, 109
  Kharkov 1941 visit, 115
  Landau, 109
  Leipunsky, 108–109, 112, 117
  letters of defence, 117
  letter to Beria, 112
  Martians comment, 108
  Müller, 120
  Pauli, 107, 109, 120
  Peierls, 107
  physicists arrested, 109
  plutonium report, 115
  Polanyi, 119
  prison experiences, 113
  Soviet Journal of Physics, 109
  SPSL grant, 110
  Thirring, 114
  University of Berne, 120
  University of Göttingen, 106
  UPTI in Kharkov, 109
  Ursell letter, 117
  von Ardenne, 114
  Weissberg, 109, 112–114
  Wigner anonymous letter, 115,
    118
Huang, Rayson, 51
Hurwitz, Charlotte, 151
Huxley, Julian, 183, 207

I
I.G. Farben, 5, 74–75, 186, 195
Ilfochrome, 80
Immerwahr-Haber, Clara, 54
Imperial Chemical Industries, 6, 47,
  184, 187

Imperial College London, 56, 184, 187, 217
Institut Pasteur in Paris, 190–191, 198, 200
Isle of Man, 128, 192, 216, 225, 228

J

Jablonec nad Nisou, *See* Gablonz
James, Edwin, 225
Johnson, Alvin, 192
Joliot-Curie, Irène and Frédéric, 112
Journal of Chemical Physics, 56, 91
Jung, Carl, 230

K

Kahn, Margarete, 151
Kaiser-Wilhelm-Institut, 12, 45, 47, 49, 54, 83, 87, 166, 195
Kalevi-Liiva, 157
Kapitsa, Pyotr, 33, 109, 112, 116
Karelia, 204
Katz, Bernard, 153
Keilin, David, 183, 188–189, 195, 198–199
Kekulé, August, 74
King Christian X of Denmark, 234
King Edward VII, 160
King George V, 184
King's College London, 183
Kornfeld, Gertrud, 54–57
Krailling, 161–163
Kral, Adalbert, 179–180, 233
Krebs, Hans, 153, 181
Kristallnacht, 20, 37, 62–63, 67, 140, 218, 233
Kuhn, Richard, 87, 89
Kurfürstenstrasse, Berlin, 63, 155, 218

L

Ladenburg, Rudolf, 47, 70
Landau, Lev, 109, 112–113, 117
Lander, Cecil, 217
Lange, Bruno, 48
Lasareff, Vladimir
    AAC application, 81, 83
    birth, 82
    Breendonk, 84–87
    Einstein friendship, 83
    Einstein letters, 83, 85–86
    Einstein reference, 83
    Freundlich, 83
    Goldfinger and Rosen, 80
    Henri, 83–85
    Israel Neumann death, 85, 87
    Kaiser-Wilhelm-Institut, 83
    Lasareff, Isaac, 84
    postwar appointments, 86
    Schmitt, 85, 87
    Simpson letters, 85
    University of Liège, 83
    Ursell enquiries, 86
League of Nations, 222
Lederer, Emil, 232–233
Lehmann, Erich
    Annalen der Physik, 68
    birth, 68
    consultancies, 69
    Fabry, 70
    Jacoby, Margarethe, 70
    Ladenburg, 68
    Lehmann, Victor, 69–70
    Miethe, 68–69
    Simpson letters, 70
    SPSL application, 69
    suicide, 70
    Technical University of Berlin, 69

University of Berlin, 68
Ursell enquiries, 71
Weissberger, 71
World War I, 69
Lehmann-Russbüldt, Otto, 184–187
Leipunsky, Alexandr, 108–109, 112, 117
Leningrad, 17–20, 27, 30, 32–34, 44, 110, 171, 189–190, 202–203, 241
Lennard-Jones, John, 16
Lindemann, Frederick, 35–36, 51, 109
Lisbon, 22, 77–78
Lodz, 32, 37, 41, 130, 132–133, 136
Loewi, Otto, 35–36, 175
London School of Economics, 37, 162
Lonnerstädter, Paul, 151
Love, Augustus, 36
Löwner, Karl, 134, 136, 139
Ludwig Maximilian University, 130
Lustig, Alfred
  birth, 95
  death, 98
  Ehrenhaft, 95
  escape to Russia, 97
  in Soviet Army, 97
  marriage, 97
  Nisko transportation, 96, 99
  Reiss, 95, 99
  Simpson letters, 96
  SPSL application, 96
  teaching award, 98
  Thirring, 95, 99
  University of Vienna, 95–96, 98–99

Ursell enquiries, 99
Yelabuga, 97–98
Lvov, 30, 38, 44, 118

M
MacRobert, Thomas, 135
Magdalen College, Oxford, 3, 12, 34–35, 47, 142–145, 148, 225, 230
Magdalene College, Cambridge, 183, 198
Mannheim, 231–232
Mann, Thomas, 29
Mark, Herman, 56, 88
Marseilles, 19, 38, 77, 81, 192–193
Marzahn-Hellersdorf District Assembly, 54
mathematicians, 129, 150
Mathematische Annalen, 146–147
Mechelen, 77, 84–85
medical scientists, 153, 179
Meister, Karl, 184–185
Meitner, Lise, 11, 54, 58, 121, 123, 124
Melchett, Lord, 74, 186
Mendelssohn Bartholdy, Albrecht, 220
Meyerhof, Otto, 194
MI5, 184–189, 197, 200–201
Miller, Richard C., 78–79
Millikan, Robert, 32
Minsk, 18, 27, 53, 156, 227
Mordell, Louis, 130, 140–145
Mott, Nevill, 34, 126, 128
Munich Agreement, 4, 20, 101, 131, 135
Munro, J. W., 184, 187
Murmansk, 203

Murray, Gilbert, 2, 222–224, 228, 230
Murrow, Edward R., 14, 60
Mussolini, 100–103

N
Narvas, 170
Nernst, Walther, 3, 11, 15–16, 36, 55, 58, 61
Neumann, Nelly, 151
New School for Social Research, 192, 219, 232–233
Nisko, 96–97, 99
Nobel Prize for Chemistry, 15, 47, 51, 64, 87
Nobel Prize for Physics, 12, 33, 59, 106–107, 115, 230
Nobel Prize for Physiology or Medicine, 153, 172
Noether, Fritz, 109
Norman, John, 205, 208
Nuffield College, Oxford, 227
Nuremberg Trial of Doctors, 178

O
Obreimov, Ivan, 109, 118
Oliphant, Mark, 125
Oppenheimer, Robert, 107, 111

P
Palestine, 19–20, 37, 64, 102–103, 187–188, 236
Paneth, Friedrich, 50
Pares, Bernard, 211
Parsons, Dorothy, 51
Pauli, Wolfgang, 16, 88–89, 92, 107, 109, 120
Pawlovsky, Yevgeny, 188

Peierls, Rudolf, 17, 36, 107, 123–128, 188, 239
Perutz, Max, 125
Pese, Herbert
    Auschwitz, 68
    birth, 66
    Breslau, 66
    Schaefer, 66
    Schrödinger, 67
    SPSL, 67
    Ursell enquiries, 68
Pfeiffer, Paul, 74
physicists, 9, 123
Piccard, Auguste, 13–14, 37, 41
Pick, George, 131–134, 151
Planck, Max, 11–12, 36, 44, 57–58, 61, 138
Polanyi, Michael, 45, 47, 87, 108, 119, 166
Popper, Karl, 230
Pose, Herbert, 121
Pringsheim, Peter, 11, 13, 15–18, 20–21, 28–29, 36–38, 42, 44, 73, 75, 82, 92–94, 103

R
Raasiku, Estonia, 156
Rahm, Karl, 233, 236
Rayleigh, Lord, 2
Red Cross, 40–41, 54, 64, 94, 167, 176–177, 199, 231–237
Reiss, Max, 37, 95, 99
Remak, Robert
    AAC application, 138
    Auschwitz, 146
    birth, 137
    difficult character, 139
    Dutch police files, 145

eccentricities, 151
Freudenthal, 142–144
Hahn, 141
Hardy, 138
Kristallnacht arrest, 140
Magdalen College, Oxford,
142–145
mathematical economics, 146
Mordell letters, 140–145
remains in Netherlands, 145
Remak, Ernst, 137
Remak, Hertha, 137–138, 140,
145
Schur, 140
Simpson letters, 141
Thomson letters, 143
University of Berlin, 138
visa, 142
Westerbork camp, 145
Remscheid, 161
Rheinheimer, Alfred
AAC application, 216
Auschwitz, 218
birth, 215
Glasgow, 216
Hamburg, 216
interned in WWI, 216
Lander, 217
Munich education, 215
Rheinheimer, Helene, 215
Sachsenhausen, 218
Stolpersteine, 218
Rhodes House, Oxford, 223
Ribbentrop, Joachim von, 35, 114
Riehl, Nikolaus, 121
Riewe, Karl-Heinrich
AAC application, 104
birth, 103

British Foreign Office, 122
CIA report, 120
Elektrostal Laboratory, 121
freed to West Germany, 122
German Physical Society, 103,
122
Houtermans paper, 105, 120
Moscow, 121
papers, 105
Soviet prison sentence, 121
SPSL letter, 105
Technische Hochschule, Berlin,
103
von Ardenne, 105
von Laue, 103–104, 114, 122
wartime research in Berlin, 121
Rintoul, William, 47, 184
Robinson, Robert, 47–51
Rockefeller Foundation, 6, 190,
219
Roosevelt, Eleanor, 113–114
Rosen, Boris
AAC application, 81, 92
Belgium, 92
birth, 92
death, 95
Franck, 93–94
Haber, 92
Henri, 93
Lasareff and Goldfinger, 80
Pringsheim, 92, 94
Simpson letters, 93
Swings, 94
University of Liège, 93–94
Ursell enquiries, 94
wartime activities, 94
Rossel, Maurice, 234–236
Röthke, Heinz, 25

Royal Society, 4, 48, 125, 130, 195,
206–207
Royal Society of Medicine, 168, 229
Russell, Bertrand, 230
Rust, Bernhard, 148
Rutherford, Lord, 2, 33, 108,
208–209

S

Sachs, Bernard, 155–156
Sachsenhausen, 67, 140, 142, 215,
218
Sachs, Paul, 211
Salz, Arthur, 236
Sante Fe, 123, 126
Sapozhnikov, Vasily, 202
Schaefer, Clemens, 66–67
Schiemann, Elisabeth, 54
Schrödinger, Erwin, 3
    Breslau, 67
    Dublin, 36
    Duschinsky, 12, 18
    Edinburgh, 36
    Einstein letter, 209
    Eisler 229–230
    Fröhlich, Herbert, 34
    Graz, 35
    letter from Duschinsky, 12
    Lindemann, 51
    Magdalen College, 35
    Pese, 67
    Princeton, 14
Schur, Issai, 140
Scuola Normale Superiore, 99
Segre, Beniamino, 101
Semon, Felix, 160–162
Sherrington, Charles, 2
Shoah Wall of Names, 29, 40

Shtepa, Konstantin, 115
Simon, Francis, 51, 55
Simons, Arthur
    AAC application, 155
    Barraquer–Simons syndrome,
        155
    birth, 155
    Cairns, 155
    Charité Hospital, Berlin, 155
    death, 157
    deportation to Raasiku, 156
    Gestapo arrest, 156
    medical movies, 155
    Simons, Hellmuth, 182
    Simpson letters, 156
    Ursell letter, 156
Simons, Hellmuth
    AAC application, 183
    arrests in France, 193
    Biological Abstracts, 192, 195,
        199
    biological warfare, 184, 186,
        195–196
    birth, 182
    Camp-des-Milles, 192
    communist suspicions, 201
    Dulles, 195
    Einstein reference, 183
    Emergency Committee, 188
    FBI report, 200
    Guggenheim, Alis, 196–197
    Hoffmann La Roche, 188,
        193–195
    Huxley, 183
    Imperial College London, 184
    Institut Pasteur Paris, 190–191,
        198, 200
    interned in France, 191–192

Keilin, 183, 188–189, 195,
    198–199
  last biomedical paper, 201
  Lehmann-Russbüldt, 184–187
  Liepmann, 185
  Marseilles, 192–193
  MI5 observations, 184–189,
    197, 200–201
  Muralt, 194
  Pawlovsky, 188
  postwar press release, 197
  Simons, Arthur, 182, 194
  Simons, Gerda, 190–192, 194,
    201
  Simons, Gertrud, 190, 194
  Simonsiella, 183
  Simons, Werner, 190–191, 194,
    196, 201
  Simpson letters, 191
  SS list, 193
  Switzerland escape, 193
  Tropical Institute, Leningrad, 190
  Ursell letters, 198
  USA move, 199
  Weizmann, 183, 188
  Wickham-Steed, 184, 195
Simonsiella, 183
Simpson, Tess, 2–3, 16–17, 27, 33,
  35, 49–51, 70, 85, 90–96, 102,
  120, 125–128, 132–135, 141,
  156, 159–160, 170, 173, 176,
  191, 226, 236
Sittig, Otto
  Auschwitz, 160
  birth, 157
  Carmichael comments, 159
  German University in Prague,
    157

Guttmann, 158
  Simpson letters, 159–160
  Sittig, Irma, 159–160
  Sittig, Trude, 159
  SPSL application, 158
  Theresienstadt deportation,
    160
  visa, 159
social scientists, 219, 221
Söllner, Karl, 47
Solovetski camp, 203
Sommerfeld, Arnold, 107, 147
Soviet Gulag, 202, 204, 121
Speer, Albert, 177
Stalin, 33, 41, 109, 112, 121–122,
  213, 234
St. Cyprien, 28, 38, 75–78
Sterbebücher Death Book, 26
Stolpersteine, 218
St. Petersburg, 82, 92
Strassmann, Reinhold, 151
Struik, Dirk, 132
Sudetenland, 20, 26
Swings, Pol, 93–94
Szilárd, Leo, 2, 108

T
Tauber, Alfred, 151
Tchernavin, Tatiana
  AAC letter, 204
  British nationality, 213
  Constable, 210
  death, 213
  escape from Gulag, 204
  Escape from the Soviets, 210
  father, 202
  Hermitage curator, 203
  In Which We Serve, 212

marriage, 202
proposed lecture tour, 210
Tchernavin, Vladimir
    AAC application, 204
    arrested, 203
    birth, 202
    Edinburgh, 211
    Einstein and Schrödinger,
        209
    escape from Gulag, 204
    expeditions in Russia, 202
    Huxley, 207
    I Speak for the Silent, 210
    Natural History - British
        Museum, 205
    Nature paper, 212
    OGPU, 203
    Salmonidae, 202
    Solovetski camp, 203
    St. Petersburg, 203
    suicide, 212
    Tchernavin, Andrei, 202, 204,
        213
    Tchernavin, Tatiana, 202–205,
        210, 213
    The Times letter, 209
    Uvarov, 206
Technical University in Munich,
    215
Teller, Edward, 108
Terenin, Alexander, 17–18
Tess, Simpson, 226–227
Theresienstadt, 26, 65, 119, 146,
    149–150, 156, 160, 170–171,
    176–179, 231–233, 237, 242
Thiessen, Peter Adolf, 47
Thirring, Hans, 33, 95–96, 99, 114

Thomson, David Cleghorn, 37, 135,
    143, 148, 155, 174, 224
Thomson, J. J., 2
Trinity College, Cambridge, 129–
    130, 141
Tyndall, Arthur, 34

U
University College London, 14, 47,
    172
University of Aachen, 147
University of Amsterdam, 149
University of Berlin, 11, 16, 44, 52,
    54–55, 58–60, 68, 82, 92, 104,
    137–138, 155, 166, 232
University of Berne, 120
University of Cambridge, 3, 16, 126,
    140, 236
University of Catania, 74, 99
University of Durham, 50
University of Heidelberg, 231
University of Leeds, 225
University of Liège, 80–89, 92–94
University of Oxford, 4, 35, 134,
    224, 230
University of Pennsylvania, 192,
    194–195, 199
University of Strasbourg, 39, 42–43
University of Utrecht, 149
University of Vienna, 31, 44, 56,
    95–96, 98–99, 106, 114, 150,
    222, 229
Ursell, Ilse, 27–29, 42, 54, 65, 68,
    71, 73, 86, 90, 99, 117, 120, 126–
    127, 156, 198–199, 220, 228, 236
US Jewish Transmigration Bureau, 38
Uvarov, Boris, 206

V
Vavilov, Sergei, 17–18
Veblen, Oswald, 131
Volta, Luigi, 102–103
von Ardenne, Manfred, 105, 114–115, 121
von Laue, Max, 11, 36, 58, 61, 66, 103–104, 114, 122
von Mises, Richard, 138
von Muralt, Alexander, 194
von Neumann, John, 5, 108, 136
von Pausinger, Franz, 222

W
Waldecker, Ludwig, 220
Warsaw, 18
Wasser, Emanuel
    AAC application, 32
    Anschluss, 34
    Auschwitz, 39, 44
    Belgium, 38
    birth, 30
    Blackett, 33
    Born, 32
    Brussels, 37
    Camp-des-Milles, 38
    Clermont-Ferrand, 39, 41–42, 194
    Dirac, 34, 42
    Drancy, 39–40, 44
    Duschinsky, 42
    Ehrenhaft, 31–33, 36–37, 42
    Einstein letter, 32
    Freedel, 38
    Fröhlich, Herbert, 34
    Gestapo arrest, 39
    Ioffe, 33

Kristallnacht, 37
Leningrad, 32–34, 44
Piccard, 37, 41
Pringsheim, 36–38, 42, 44
Shoah Memorial, 40
SPSL letters, 37
St. Cyprien, 38
University of Strasbourg, 39, 42–43
University of Vienna, 31, 44
Ursell enquiries, 42
Wasser, Anna, 30
Wasser, Daniel, 30
Wasser, Elisabeth, 37–38, 40–41
Wasser, Sara, 32, 37, 40–42
Zeitschrift für Physik papers, 31
Weinmann, Erwin, 233
Weissberg, Alexander, 109, 113
Weissenberg, Karl, 15
Weizmann, Chaim, 61, 89, 183, 187–188
Wells, H. G., 112
Westerbork, 145, 149, 235
Weyl, Hermann, 150
Whitehead, J. H. C., 65, 129, 131, 134, 148
Whittaker, Edmund, 136
Wickham-Steed, Henry, 184, 195
Wigner, Eugene, 5, 108, 114–115
Wikkenhauser, Gustav, 105
Windaus, Alfred, 64
Wreschner, Marie
    Abderhalden, 48
    birth, 44
    British Federation of University Women, 50, 54
    Donnan, 47, 49

Freundlich, 45, 47, 49–52
Fröhlich, Cecilie, 54
Haber, 45, 47
handbuch, 48
Hinshelwood, 51
Kornfeld, 54–57
Lange, 48
Lindemann, 51
Meitner, 54
mother death, 52
Paneth, 50
Parsons, 51
patent, 45
Photoelectric Cells report, 48
Robinson, 47–51
Simpson letters, 49

SPSL application, 48
street naming, 54
suicide, 53
thesis, 44
transport to Minsk, 52
Ursell enquiries, 54
Wehnelt, 45
Wreschner, Jacob, 44

Y
Yelabuga, 97

Z
Zurich, 17, 41, 57, 81, 135, 164, 194–195
Zylberszac, Anna, 41–42